The Joyce Well Site

On the Frontier of the Casas Grandes World

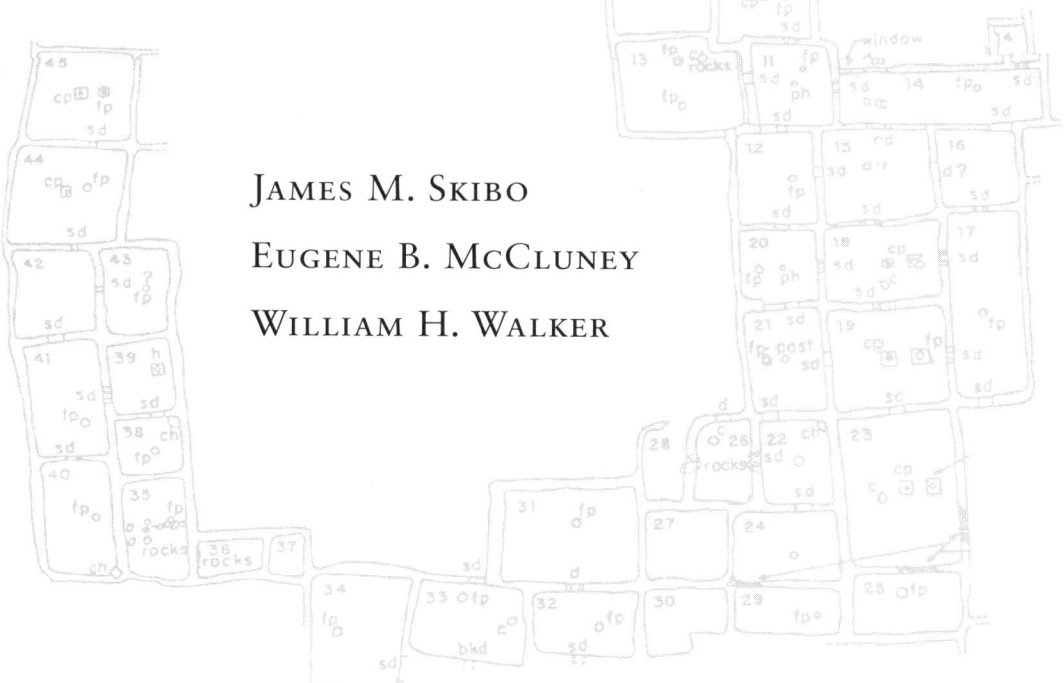

James M. Skibo

Eugene B. McCluney

William H. Walker

The University of Utah Press

Salt Lake City

© 2002 by The University of Utah Press

07 06 05 04 03 02
5 4 3 2 1

The Defiance House Man colophon is a registered trademark of
The University of Utah Press. It is based upon a four-foot-tall
Ancient Puebloan pictograph (late PIII) near Glen Canyon, Utah.

LIBRARY OF CONGRESS CATALOGING-IN-PUBLICATION DATA

Skibo, James M.
    The Joyce Well Site : on the frontier of the Casas Grandes world /
James M. Skibo, Eugene B. McCluney, William H. Walker.
        p. cm.
Includes bibliographical references and index.
    ISBN 0-87480-728-X (pbk. : alk. paper)
    1. Joyce Well Site (N.M.) 2. Casas Grandes culture—New Mexico—
Hidalgo County. 3. Excavations (Archaeology)—New Mexico— Hidalgo
County. 4. Archaeological dating—New Mexico—Hidalgo County.
5. Magnetic resonance—New Mexico—Hidalgo County.
6. Plant remains—New Mexico—Hidalgo County. 7. Human remains
(Archaeology)—New Mexico—Hidalgo County. 8. Hidalgo County
(N.M.)—Antiquities. I. McCluney, Eugene B., 1928– II. Walker,
William H., 1964– III. Title.
    E99.C23 S55 2002
    978.9'693—dc21
    2002004652

Gene McCluney dedicates this book
to his son Mark and daughter Erinn
who know the wonders of the great outdoors.
And to Richard Sense, foreman at Joyce Well,
his everlasting thanks.

William Walker dedicates this work
to John M. Kipp, a good friend.

James Skibo dedicates this book
to Becky, Matt, and Sadie.

CONTENTS

FIGURE A.1. The 1963 Joyce Well excavation team. Bottom row (left to right): George O'Brien and Richard Barkley; top row (left to right): Clem Jackson, Eugene McCluney, Ray Deuser, Resse Upshaw, Russell Baker, Richard Sense, and Bob Baker. Not pictured: Dick Fike.

ACKNOWLEDGMENTS

The survey, excavation, and reporting of the Joyce Well site was made possible by many individuals. McCluney would like to extend lasting appreciation to the following institutions and individuals. The Executive Committee of the School of American Research made this project possible by their generous funding of Hidalgo #3. I would like to thank the officers and ranchers of the Diamond-A Ranch, Hidalgo County, New Mexico, for their many courtesies extended to myself as well as my crew. Special thanks to Mr. and Mrs. George Upshaw of the Culberson Ranch, and Mr. and Mrs. V. F. Tannich of Deming, New Mexico, for their help in many ways and for their constant interest during the excavation of the Joyce Well site. Dr. Charles Di Peso and his staff of the Amerind Foundation, Dragoon, Arizona, assisted me by furnishing suggestions and study material from the Casas Grandes site. I would also like to thank Dr. Erik Reed, National Park Service, for his analysis of the skeletal material, and Dr. Hugh Cutler, Missouri Botanical Gardens, St. Louis, Missouri, for the analysis of the plant remains. Mrs. Phyllis Hughs of Santa Fe, New Mexico, made many of the drawings found in Chapter 2. Last, but not least, I wish to extend my appreciation to the field crew of Hidalgo #3 Project (Fig. A.1), who devoted days and nights of hard work to contribute to the information contained in this report. Without them this report would not have been written. The 1963 field crew consisted of Dick Sense, foreman, Bob Baker, Russell Baker, Richard Barkley, Ray Deuser, Dick Fike, Clem Jackson, and George O'Bryan. If any individual has been overlooked in these acknowledgments, I wish to state that the oversight is mine and additional appreciation and apology is here made.

Skibo and Walker would like to acknowledge that this work was supported by the National Science Foundation (BCS-00014322) and a variety of small grants from Illinois State University and New Mexico State University. The 1999 and 2000 excavations were done as field schools and we would like to thank our department chairs, Nick Maroules, Illinois State University, and Scott Rushforth, New Mexico

FIGURE A.2. The 1999 Joyce Well excavation team. Bottom row (left to right): Beth Bickel, Andrew Briik, Richard Raffaelli, Lonnie Ludeman, Tom Todsen, Julie Jeakle, Becky Skibo, and James Skibo; middle row (left to right): William Walker, Christa Clement, Tony Eckert, Laura Ogawa, Sadie Skibo, Marianna Sachese, Amanda Donato, Lee Panich, Louis Armstrong, Lane Howard, Venessa Joseph, Caroyl Scott, Janice Greenslade, and Jim Pisell; back row (left to right): Mike LaVerde, Larry McBride, Margorie Fancher, Matt Skibo, Gaea McGahee, and Mike Hanely. Not pictured: Dave Brown and Steve Eckburg.

State University, for their support. Mickey McCombs coordinated the financial aspect of the field work and JoAnne Geigner helped out considerably in the preparation of the 1963 manuscript. We would also like to thank the staff and field crew for their hard work and Gabriela Soria and Becky Skibo for their support (Fig. A.2). This work was made possible by the Animas Foundation, directed by Ben Brown, and the Gray Ranch, current owners of the site. Ben Brown and Seth Hadley, ranch owner and foundation board member, were especially supportive and helpful in providing logistical support to our field projects. Ben deserves special thanks for making our project possible. Other staff members of the Animas Foundation were also very helpful. Gerald Malzac, carpenter and handyman extraordinaire, made our stay much more comfortable. The Animas Foundation office staff, Kelly Peterson and Jennifer Medina, never complained about the extra burdens of having 30 archaeologists descend on them each summer, and Sam Smith, fire manager, Ric Meloy, windmill man, and Mary Moore, manager of ranch security, all helped to make our work possible. John and Rex Kipp also helped this project get off the ground. We also appreciate the comments of Steve Lekson, Dave Phillips, and one anonymous reviewer. Special thanks to Jeff Grathwohl, director, and the staff of the University of Utah Press for their usual fine work.

ONE

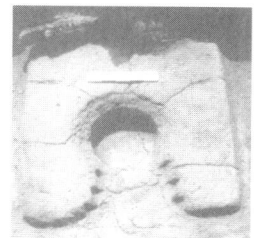

Joyce Well and the Casas Grandes World

WILLIAM H. WALKER, JAMES M. SKIBO,

AND EUGENE B. MCCLUNEY

The Joyce Well site is located in the extreme southwest corner of New Mexico, in an area referred to as the "Bootheel" (Fig. 1.1). The site lies along the middle reaches of Deer Creek, a north-south trending intermittent stream that begins in the hills east of Animas Peak and drains into the Playas Valley of southwest New Mexico. The pueblo contains an estimated two hundred rooms (Fig. 1.2) and is one of several large Animas phase villages (A.D.1200–1400) found in the Bootheel region comprised of the Playas, Animas, and San Bernadino valleys of southwest New Mexico and southeast Arizona (Kidder et al. 1949). Located approximately 10.5 km north of the border between the United States and the Republic of Mexico, Joyce Well holds critical information for understanding variability in the Animas phase. Joyce Well and the Animas phase sites more generally represent a northern frontier of the Chihuahuan culture horizon centered on the Casas Grandes Valley and the well-known site of Paquimé, also known as Casas Grandes. Two interrelated research questions have inspired the research compiled in this volume: What is the nature of the Animas phase as represented at Joyce Well and how does that data contribute to a larger understanding of the role Paquimé played in the international four corners region?

Communities in this regional system, including Joyce Well, possessed similar artifacts (e.g., shared polychrome-style ceramics, scooped metates), architecture (e.g., adobe pueblos, T-shaped doors, platform hearths, ball courts), and rock art (e.g., feathered serpents, cartouches) indicative of a reservoir of shared beliefs and practices (e.g., Di Peso

FIGURE 1.1. Location
of the Joyce Well site
and its relationship to
Casas Grandes.

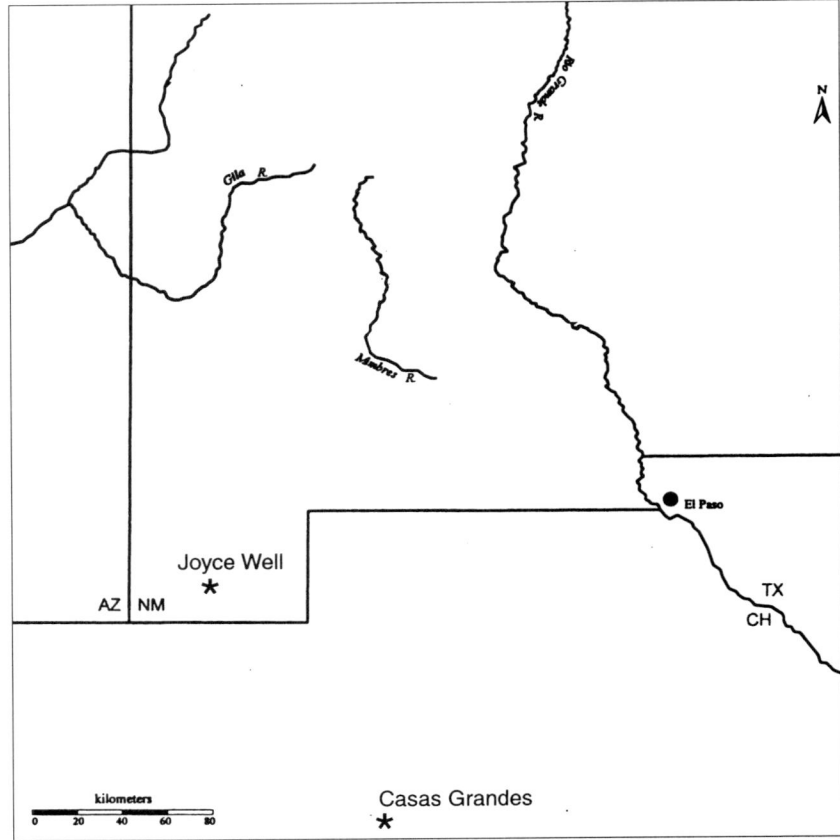

FIGURE 1.1. Location of the Joyce Well site and its relationship to Casas Grandes.

1974; Ravesloot 1979; P. Schaafsma 1998). The organization of interaction in the region, however, remains murky because until recently the majority of research has focused on the center at Casas Grandes. It is still unclear how ritual, economic activity, or political leadership was organized in the various outlying communities of the region. Although Casas Grandes is relatively well known (Di Peso 1974; Di Peso et al. 1974), the majority of sites remain untested or unreported (Schaafsma and Riley 1999:9; Whalen and Minnis 1999:36). Until more detailed excavation and publication occurs it will be difficult to understand the processes that shaped the Casas Grandes world between A.D. 1200 and 1400. To counter that trend, this volume brings together previously unpublished work from Joyce Well.

Walker and Skibo initiated background research in 1997 to write a National Science Foundation grant for study at Joyce Well and realized that a large amount of useful research had already been done but for various reasons had not been published. This research covers forty years beginning with a large-scale excavation of Joyce Well by the School of American Research under the direction of Eugene McCluney (1960s). At Curt Schaafsma's suggestion we contacted McCluney and he gener-

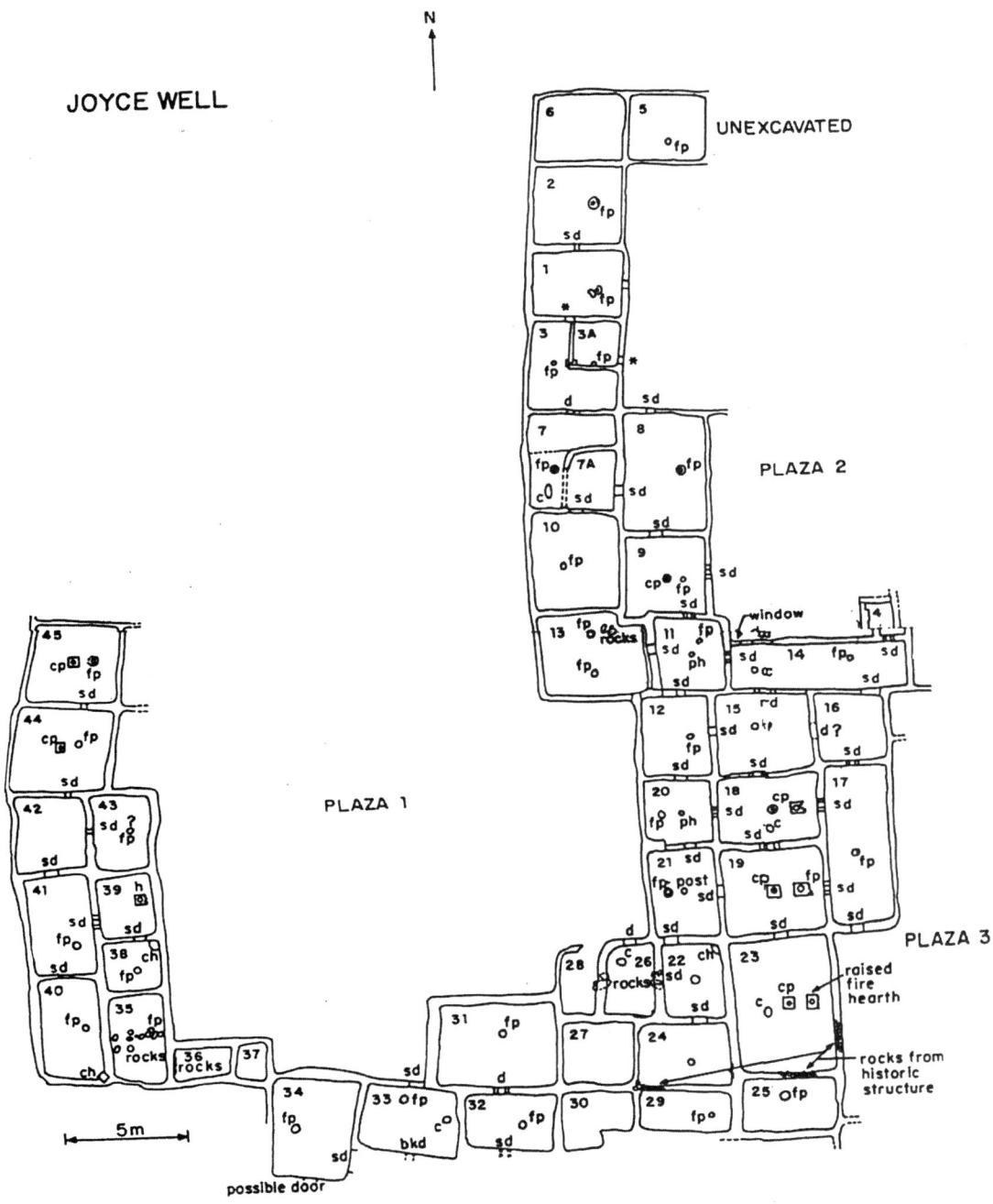

FIGURE 1.2. Rooms excavated during the 1963 field season. A detailed drawing of each room can be found in Chapter 2.

ously agreed to collaborate with us in our new research project. We agreed to prioritize the publication of McCluney's original field report as well as other unpublished work from the site including results from John Carpenter's master's proposal (1980s), and specialized studies initiated by Curtis Schaafsma, former curator of archaeology and state archaeologist of the Museum of New Mexico (1980s and 1990s). Our first year's study of the Joyce Well ball court was also included.

THE BOOTHEEL AND THE CHIHUAHUAN CULTURE HORIZON

The earliest professional excavations in the Bootheel were conducted in the Animas Valley at the Pendleton Ruin with the goal of discovering the relationship between the "little-known Casas Grandes culture" and the peoples of southern New Mexico and Arizona (Kidder et al. 1949:109). Through ceramic cross-dating, Kidder and others (1949:147) placed the Pendleton Ruin in the mid fourteenth century. Contrary to expectations, many of the diagnostic traits associated with the Casas Grandes core area, such as T-shaped doorways, platform hearths, under-floor burials, and deeply scooped metates, were absent (Kidder et al. 1949:144). Although the inhabitants of the Pendleton Ruin constructed adobe room blocks around large plazas and used Chihuahuan-style pottery, they appeared to be "country cousins" at best. To distinguish these frontier people from others living closer to Casas Grandes, the fourteenth-century pueblos of the Bootheel were designated the Animas phase. Di Peso's (1951, 1953, 1956, 1968, 1974; Di Peso et al. 1974) subsequent archaeological, ethnographic, and ethnohistorical study of Casas Grandes and other sites in the Southwest revised these findings in several insightful, albeit controversial, political and economic interpretations of the center and its relation to peripheral regions such as the Bootheel.

Di Peso and other Southwestern archaeologists, in common with Caldwell (1964), opposed isolationist models of culture history that left macroregional interactions unexamined (Doyel 1994). As an alternative, they proposed core-periphery models driven by trade and politics to describe relations between Casas Grandes and its hinterlands, and more broadly, the North American Southwest and Mesoamerica (e.g., Di Peso 1974, 1983; Kelley 1966; Whitecotton and Pailes 1986). Di Peso envisioned the large towns of Casas Grandes, Chihuahua (Di Peso 1983), and Pueblo Bonito, New Mexico, as exploitative mercantile centers that channeled turquoise and other exotic raw materials to southern Mexican states (e.g., Tula). Although ultimately driven by economic interests, the stability, or perhaps instability, of this economic organization depended upon a combination of coercive force (warriors) and religious ideology (the *Quetzalcoatl* cult). Di Peso's (1968) appreciation for the long-term effects of trade on this region's archaeological record, similar to Caldwell's (1964) recognition of the integrative power of shared rituals and ritual artifacts, provided an exciting alternative to previous explanations (diffusion, migration, and invention) of archaeological variability in the North American Southwest.

According to Di Peso, Casas Grandes was a thirteenth-century frontier city analogous to ethnohistoric Aztec and Tarascan enclaves founded by traveling merchants, *pochteca*. Presumably the political, economic, and religious hegemony these merchants cast over the northern frontier, including the Bootheel, resulted in the material culture patterns of the Chihuahuan culture (e.g., Brand 1935; Carey 1931; Lister 1946) and its northern frontier areas now defined by the Animas (Kidder et al. 1949), Black Mountain (LeBlanc 1980a), and El Paso phases (Lehmer 1948). At Casas Grandes, these entrepreneurs centralized the production and distribution of shell (Di Peso et al. 1974:4:424–430, 437–439, 5:661–665, 8:passim), copper (Di Peso et al. 1974:8:passim), pottery (Di Peso et al. 1974:4:437–439), macaws (Di Peso et al. 1974:4:389, 394, 436, 5:478–482, 491, 506–514, 528–542, 801–802, 818–820), and turkeys (Di Peso et al. 1974:4:383–384, 5:584–592). Their organizational skills, coercive power, and knowledge also made possible the construction of platform pyramids (Di Peso et al. 1974:4:305–316, 465–468, 471–474), ball courts (Di Peso et al. 1974:5:617–620), effigy mounds (Di Peso et al. 1974:5:475–478), ceremonial rooms (Di Peso et al. 1974:4:407–410, 460–462, 5:506–514, 553, 593, 647–650, 763), civic water works (Di Peso et al. 1974:4:375–381, 5:830–853), and an agricultural infrastructure spanning the entire Casas Grandes Valley (Di Peso et al. 1974:5:825). In short, Mesoamerican *pochteca* turned a previously egalitarian Mogollon farming village into a socially stratified urban outpost. In the A.D. 1340s the city, already in decline, suffered a fiery destruction at the end of the appropriately named Diablo phase (Di Peso et al. 1974:4:205). Invaders, perhaps from the west, stormed the city and destroyed it during what Di Peso termed the "Chichimecan revolt" (Di Peso 1974:2:320, 3:758). Subsequent studies of Southwestern warfare and violence (LeBlanc 1999; Turner and Turner 1998) continue to promote this version of Chihuahuan culture history (see Walker and Skibo, Chapter 8, this volume).

Following closely on the heels of Di Peso's work in Chihuahua, McCluney (Chapter 2, this volume) initiated new excavations in the Bootheel at Joyce Well, Clanton Draw, and Box Canyon sites to assess Di Peso's political and economic hypotheses. In contrast to the Pendleton data, these villages contained more obvious Casas Grandes traits, including under-floor burials, platform hearths, and Chihuahuan doorways (Fig. 1.3). However, significant frequencies of Chihuahuan polychromes recovered from these sites were locally manufactured (see McCluney, Chapter 2; Carpenter, Chapter 7, this volume). McCluney (1965a:40) tentatively inferred from these data that the Bootheel may have been tied to Casas Grandes through either an ethnic (migration) or an undefined political relationship. At the time of these excavations, inferences of ritual abandonment activities were rare. As a result, stratigraphic variables critical for determining the causes of burning and other site formation processes (sensu Schiffer 1987) at either Joyce Well or Casas Grandes were not systematically recorded. Instead, attention

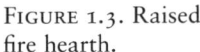

FIGURE 1.3. Raised
fire hearth.

was focused on the study of chronology and the political and economic consequences of artifact distributions.

Di Peso's early dating of the Chihuahua culture area dramatically shaped his interpretation of these Joyce Well data. Relying upon several uncalibrated radiocarbon dates (A.D. 1565, 1595, 1620) recovered by McCluney, Di Peso (1974:3:778, 970, n. 72) explained the architectural and artifactual evidence at Joyce Well (and the Bootheel, more generally) as the product of a refugee population fleeing Casas Grandes after the Chichimecan revolt (Di Peso et al. 1974:4:125). DeAtley's (1980:72–73) subsequent calibration of these anomalous radiocarbon samples, however, demonstrated that Joyce Well actually dates to the mid fourteenth century, contemporary with Casas Grandes (Schaafsma et al., Chapter 6, this volume). Subsequent redating of Casas Grandes (Dean and Ravesloot 1993) to between A.D. 1200 and 1450 corroborates these findings for the horizon more generally. Moreover, her analysis of local variants of Ramos Polychrome indicated that fine decorative differences distinguished these locally made vessels from Ramos Polychrome recovered from Casas Grandes (DeAtley 1980:143). These data led her to conclude that neither extensive trade nor a late migration of Casas Grandes peoples (potters) could account for the presence of these local Ramos-style ceramics. Woosley and Olinger (1993) later confirmed these results identifying locally made Ramos pottery through NAA analysis. Recent archaeomagnetic sampling of hearths from Joyce Well (Schaafsma et al., Chapter 6, this volume) confirms its relative contemporaneity with Casas Grandes.

During the last 20 years of excavation and survey in the Bootheel (Lambert and Ambler 1965; Brand 1943; DeAtley 1980; Duran 1992;

Findlow and DeAtley 1978) and other parts of the greater Chihuahuan region (Creel 1997; Minnis 1988; Ravesloot 1979; Schaafsma 1995; Whalen and Minnis 1996, 2001), a new historical trajectory has emerged for the Chihuahuan culture that contrasts sharply with Di Peso's interpretation. These studies have undermined most facets of Di Peso's macroregional interpretation (Braniff 1986; Dean and Ravesloot 1993:103; DeAtley 1980:158; Douglas 1995; Minnis 1984, 1988; P. Schaafsma 1997; VanPool et al. 2000; Whalen and Minnis 2001; Woosley and Olinger 1993), and suggest that in the fourteenth century the town of Casas Grandes shared ritual imagery and activities with its periphery, but did not dominate it economically or politically. Minnis (1984, 1988), in particular, found that shell, copper, and parrots, all exotic objects that figure prominently in Di Peso's model, were produced and hoarded for use at Casas Grandes rather than for distribution in a large regional trade network under the town's supposed control. Although some of these exotics did go north, and turquoise south, trade in these exotics does not appear to have fueled the rise of Casas Grandes. Instead Casas Grandes may have been the primary consumer of these artifacts.

OUTLINE OF THE VOLUME

McCluney's report describing his excavation of Joyce Well forms the core of the volume (Chapter 2). He wrote the bulk of this report shortly after the completion of his 1963 field season. Although the report went through a number of revisions in the intervening years, it remained unpublished until now. Mimeograph copies and later photocopies of this report have been circulated and cited by Southwestern archaeologists since the mid-1960s. Its final publication is particularly important given that it may contain the best descriptions of unlooted rooms at the site. Local folklore states that the entire pueblo was systematically looted at the end of McCluney's field season.

The final revision took place in 2000 when Skibo and McCluney prepared the 1963 material for publication. The original report by McCluney has been changed in two significant areas. First, edited versions of the room excavation and burial forms are included in this chapter. Some minor details are omitted but the original forms are available at the Archeological Records Management Section, Historic Preservation Division, housed at the Laboratory of Anthropology, Museum of New Mexico, in Santa Fe should anyone find it necessary to review the original documents. Second, the drawings of each room, which include important architectural information, have also been added.

The presentation of the research spanning over thirty years has been a challenge. McCluney completed his excavations and initial report prior to Charles Di Peso's formulation of a comprehensive model of the Chihuahua culture (1974). In the intervening years that model has served as the focal point of much of the research in the region and has come under significant revision in the last decade. The same thirty years has also witnessed a revolution in archaeological method and theory and attendant interpretations of archaeological evidence. Rather than attempting to update his report to address these changes, McCluney suggested that it would be more useful to publish it with minor revisions in its original form. This report was based on techniques and assumptions from another time and masking those would not change his original observations but instead delay their publication. The revisions instead revolve around the organization of the data presented. Mc-Cluney's original report resembles the Pendleton report (Kidder et al. 1949) and other culture history studies from the Southwest.

Shortly after the 1963 excavation, McCluney had the skeletal remains analyzed by Eric Reed and the botanical data analyzed by Hugh and William Cutler. These were originally projected to be appendices in a site report but we have made them short chapters in this volume. Chapter 3, Plant Remains from the Joyce Well Site, was written by Hugh and William Cutler. They identified corn, several varieties of gourd, and cotton.

A total of 27 subfloor burials were removed during the 1963 excavation. In all cases the remains were in a poor state of preservation. The remains from 23 burials were in good enough condition to send to Eric Reed for further study, and his results are reported in Chapter 4. Reed's report is followed by Skibo and Walker's (Chapter 5) description and analysis of the Joyce Well ball court. This court, which is one of three within 11 km of Joyce Well, is an important link to Casas Grandes. The shared material culture traits, especially those linked to ritual activities such as ball courts, dance plazas, and rock art, indicate some form of prehistoric religious integration of ritual activities and beliefs (Ravesloot 1979:89; Schaafsma 1997; Wilcox 1995; see Skibo and Walker, Chapter 3, this volume), perhaps involving pilgrimage activities (Fish and Fish 1999:23–24; Walker and Skibo, Chapter 8, this volume). This integration, however, does not appear to involve a regionally powerful priesthood, but instead local communities of various sizes with varying forms of religious leadership. The peoples of the Bootheel employed similar icons and ritual buildings (plazas and ball courts) in their own local ritual tradition that never attained the complexity of Casas Grandes.

In a recent survey of the interaction sphere, Whalen and Minnis (1996, 1999, 2001) defined three concentric zones surrounding Casas Grandes differentiated in part by such ceremonial variability. In the inner zone large towns such as Casas Grandes possess formal ball courts and multistory adobe architecture. The second zone consists of smaller villages without identifiable ball courts, albeit other architecture and artifacts are similar to those found in the inner zone. The outer zone includes the Bootheel. In this region ball courts occur but they are shorter and broader than those of the inner zones. Although artifacts and architecture are also similar, the differences would impact use and performance. We explore a performance-based analysis of this socially integrative feature suggesting it was used in a fertility-oriented ritual game.

In Chapter 6, Schaafsma, Cox, and Wolfman provide important archaeomagnetic evidence for establishing terminal occupation dates of Joyce Well and the region. In this chapter they review models of the organization of the Chihuahuan culture that have tended to balkanize the Chihuahuan culture area into independent regions associated with local phases. They suggest that more detailed chronological study of sites will allow for a more comprehensive understanding of the relations between these regions. Returning to several of the rooms originally described by McCluney, they carried out archaeomagnetic dating finding evidence of occupation spanning the thirteenth and fourteenth centuries. The stratigraphic positions of the hearths they sampled suggest an episode of room abandonment in the mid thirteenth century and a second in the mid fourteenth century. Preliminary analysis of archaeomagnetic dates from burned room walls and hearths from the 2001 season are consistent with this bimodal distribution (Blinman, pers. comm. 2001).

In Chapter 7, Carpenter explores ceramic links between Joyce Well and other regions in the Chihuahuan culture area. Although he agrees with Minnis and Whalen (1996) that the growth of Casas Grandes and surrounding regions resulted from local processes, he does not accept their concentric spatial model of cultural interaction where the most intensive relations are in the immediate area (30 km radius) around Casas Grandes. Instead he finds that the predominant decorated pottery at Joyce Well, locally manufactured Ramos Polychrome, links its population closely to people living in the Casas Grandes Valley. This is especially significant given his observation that Carretas-Huẽrigos Polychrome characterizes the decorated Chihuahuan ceramics of their more immediate neighbors to the south in the Carretas Basin. He argues that Chihuahuan polychrome ceramics (e.g., Babicora, Carretas-Huẽrigos, Villa Ahumada) have strong regional associations and puts forth the intriguing hypothesis that Ramos Polychrome correlates with the distribution of ball courts because these sites have more, albeit as yet undefined social, religious, and economic interaction with Casas Grandes.

In the final chapter, Walker and Skibo (Chapter 8) suggest a new and complimentary research direction, the study of the religious ties across the Chihuahuan culture horizon. If the peoples of the Chihuahuan culture shared a common religious organization that included some seasonal pilgrimage activities at Casas Grandes, then one could explain through ritual activities several aspects of the archaeological variability encountered throughout this region. The depiction of horned snake imagery, parrots, and abstract icons, and the use of stylistically similar architecture and artifacts (e.g., Ramos Polychrome ceramic styles, platform hearths, ball courts, and plazas) across the interaction sphere suggests a reservoir of shared beliefs and practices at the household and community scale. The largest centers, such as Casas Grandes, would attract the most ceremonial goods. The hoarding and use of shell, parrots, copper, and Gila Polychrome pottery at Casas Grandes may simply be a product of their ritual use at the center (Lekson 1999; Minnis 1984, 1988) rather than in its periphery. In such a scenario, these materials were manufactured, stored, used and often ritually discarded by a ceremonial constituency of town residents and perhaps pilgrims. We suggest that such analysis should look at not only the acquisition and use of material culture in the region but also their eventual discard. The ritual abandonment of Joyce Well, Casas Grandes, and other sites in the interaction sphere may contain clues to varying degrees of religious integration.

CONCLUSION

These studies are an invaluable foundation for ongoing research at the site. In aggregate they demonstrate that data from Joyce Well are critical for understanding the organization of the Chihuahuan culture horizon. Archaeological study of past cultures, trading networks, religious interaction spheres, and other analytical renderings of organized behavioral interaction cannot be described and explained unless their constituents parts are placed in time. The dating of Joyce Well is becoming ever more accurate and precise. Knowledge of ceramics, ritual architecture, and abandonment have also increased and suggest directions for more specialized research.

TWO

The 1963 Excavation

EUGENE B. McCLUNEY

ENVIRONMENT

REGIONAL GEOGRAPHY

The Joyce Well site is located in southern Hidalgo County, New Mexico. This region is characterized by a series of north-south oriented mountain masses, which are often grass covered, wide river valleys, and broad expanses of desert and alkali flats. The site lies on the eastern side of the Animas Mountains, which is a 38-mile-long high-altitude range that divides this region into two valleys: the Animas Valley on the west and the Playas Valley on the east. The eastern boundary of the Playas Valley is the southern tip of the Little Hatchet Range, the Big Hatchet Range, the U-Bar Ridges, and the Alamo Hueco Range.

From Hatchet Gap, which separates the Little Hatchet from the Big Hatchet ranges (Schwennesen 1918), the country emerges into a wide plain that borders upon the eastern line of Grant County, New Mexico. Paralleling either side of this plain is complicated systems of mesas and hills that are crosscut by arroyos and washes. Continuing southward, Hatchet Gap becomes the foothills of the Big Hatchet Range, which reaches a height of 8,366 ft.

The Big Hatchet Mountains tend to widen from east to west as one moves south and their orientation swings more to the southeast. Rising from the valley floor at this point are two groups of isolated high-scarped masses know as the U-Bar Ridges (Zeller 1965). The ridges form a natural barricade for a secondary valley, named the Mojado Pass, which leaves the main valley in a southeasterly direction. The U-Bar Ridge group marks the end of the Big Hatchet Mountains and gives rise to another, but smaller, mountain mass called the Alamo Hueco Range. Like the Big Hatchet Range these mountains are oriented northwest-southeast. Approximately 10 miles in length, the Alamo Huecos merge into the valley floor just north of the international border and are reduced to insignificant foothills in Chihuahua.

Looking west from Hatchet Gap across the Playas Valley is a distance of 15 to 20 miles of typical Chihuahuan Desert Grassland until the northern foothills of the Animas Mountains. Schwennesen (1918) recognized the Animas Mountains as originating in the Pyramid Peak area of Lordsburg, New Mexico, and saw a possibility that these now-separated mountain formations were once joined to form one long mountain mass. The highest elevation of the Animas Mountains is Animas Peak, 8,519 ft above sea level. On the southern and southeast end of the Animas Mountains are a series of small rises referred to as the San Luis Hills, the White Mountains, and the Hilo Mountains.

On the eastern edge of the Animas Mountains several impressive but small valleys make their way into the longer Playas Valley. Flowing in these valleys are three large creeks, the Walnut, Brushy, and Deer, that reach the Playas Valley by circuitous routes.

Joyce Well is located on Deer Creek, which has a watershed that extends from the Continental Divide that runs along the crest of the Animas Mountains. The course of the creek is oriented almost due south in the higher elevations but as it makes its way down the mountain slopes and enters into the lower and flatter elevations, it turns to the southeast and branches into two fingers called the Upper and Lower Deer Creek. Together these two creeks form the Deer Creek drainage, which is bounded by the Hilo, Whitewater, and San Louis Mountains, form an almost-hidden valley. Thus, lying southeast of the Animas Range and west of the Hachita Valley we have an isolated area cut off from direct connection with open terrain.

CLIMATE

The climate of the region is semi-arid with a mean annual precipitation of about 10 inches (Moore-Craig 1996). About 60 percent of the annual precipitation comes during the monsoon season, which begins about July 1 and lasts through September. Rainfall (and some snow), however, is quite unpredictable and can fluctuate a great deal from year to year. Mild to heavy winds are a constant in this region, which typically blow out of the southwest. Summer temperatures rarely exceed the high 90s and the highest recorded temperature is 107 degrees Fahrenheit. Winters are mild and there are fewer than 100 days of freezing temperatures (see Moore-Craig 1996 for a more complete discussion of climate).

The Deer Creek Valley presents a climatic contrast to the vast desolation found in the Hachita Valley to the east. On entering the Deer Creek area one is instantly aware of the change in the vegetation, especially the large trees and other plants on the creek borders. Although the larger conifers of the Animas Mountains to the west do not extend into the valley, there are large Arizona sycamore, oak, and juniper trees that extend the total length of both Deer and Brushy Creeks.

WILDLIFE

Our research area has tremendous biological diversity because it lies at the crossroads between the flora and fauna of the Sierra Madre Mountains to the south, the Rocky Mountains to the north, the Chihuahuan Desert to the east, and the Plains Grassland to the northeast. What is more, the Animas Mountains are one of the 40 "sky islands," which are borderland mountains with great biological diversity owing to altitudinal changes, that serve as a transitional zone between the Rocky Mountains to the north and the Madrean Archipelago to the south (see Moore-Craig 1996:15–17). Few areas in the continental U.S. can match this biodiversity of animals that include 370 species of vertebrates (252 species of birds, 71 mammals, 42 reptiles, and 11 amphibians). Table 2.1 provides a list of 71 contemporary species of mammals identified in the study area. Although there has been some change in plants and animals since Joyce Well was occupied, there has not been significant changes in the environment and wildlife habitat for the last 750 years.

Plant life is equally diverse in the study area as a result of variations in altitude and available moisture. Seven different ecosystems have been identified in the study area. These include: Montane Coniferous Forest (7,500 ft and higher), Madrean Oak-Pine Woodland (6,100 to 7,500 ft), Madrean Evergreen Woodland (5,500 to 6,100 ft and wetter), Interior Chaparral (5,500 to 8,500 and drier), Plains Grassland (5,300 to 6,000 ft), Desert Grassland (4,800 to 5,600 ft, and wetter), and Chihuahuan Desert Scrub (4,800 to 5,600 ft, and drier).

The region has tremendous biodiversity and thus there was a large range of plants and animals that could have been exploited by the prehistoric inhabitants of the region.

SITE DESCRIPTION

LOCATION

During reconnaissance survey made during the excavations of Clanton Draw and Box Canyon ruins in 1962 (McCluney 1965a), we visited the Deer Creek drainage of the Animas Mountains. Approximately 8 miles north of the Culberson Ranch, and running parallel to the Upper Deer Creek, we located a large area with much surface material (Fig. 1.2). Located near Deer Creek Well (which at the time consisted of a windmill, cement cistern, and steel tank used for watering stock), the site lies on the east bank of a deep, short valley supporting a narrow creek-bottom

TABLE 2.1. Mammals of the Gray Ranch

SPECIES	LOCATION	OCCURRENCE
Virginia opossum	R	1 record; Eicks Tank
Arizona shrew	F	2 records; Turkey (Aspen) Springs
Desert shrew	GOR	
Long-tongued bat	GOR	Typically montane
Sanborn's long-nosed bat	GR	
Cave *myotis*	GORD	
Southwestern *myotis*	OFRC	
Fringed *myotis*	GOFRDC	Most abundant in desert scrub and grassland
Long-legged *myotis*	OFRC	Intermediate elevations
California *myotis*	GOFRDC	
Small-footed *myotis*	GOFRDC	
Silver-haired bat	OFC	3 records, 25 May 81
Western pipistrel	GOFRDC	Most abundant in grassland and oak woodland
Big brown bat	GOFRDC	Most at higher elevations
Red bat	OR	Along major drainages
Hoary bat	GOFRDC	Typically montane
Southern yellow bat	OR	
Townsend's big-eared bat	OFC	
Desert pallid bat	GORD	
Brazilian free-tailed bat	GORD	
Big free-tailed bat	GD	Tank near San Luis Pass
Eastern cottontail	FC	Upper elevations
Desert cottontail	GORD	Lower elevations
Black-tailed jackrabbit	GORDC	Typically shrub-grassland
White-sided jackrabbit	G	South Animas and Playas valleys
Cliff chipmunk	OFCR	Near rocky outcrops
Harris' antelope squirrel	R	Lower Deer Creek, Double Adobe, Cottonwood Canyon, and McKinney Tank
Spotted ground squirrel	GR	Sandy soils
Rock squirrel	GOFRDC	Broken terrain, all elevations
Black-tailed prairie dog	G	Animas Valley; extirpated
Botta's pocket gopher	GORD	Below 1,675 m
Southern pocket gopher	OFC	Above 1,675 m
Silky pocket mouse	GORD	Typically open grassland
Hispid pocket mouse	GRD	Dense grasses
Rock pocket mouse	ORDC	Gravely or rocky soils
Desert pocket mouse	GRD	Fine alluvial soils at lower elevations
Ord's kangaroo rat	GRD	Typically mesquite grassland and finer soils
Banner-tailed kangaroo rat	GRD	Mesquite grassland and bajadas
Merriam's kangaroo rat	GORD	Up to 1,800 m, rockier soils
Western harvest mouse	GORD	Grassland at all elevations
Fulvous harvest mouse	R	1 record; in Rock Ridge exclosure of Deer Creek
Cactus mouse	D	Rocky, dry hillsides
Deer mouse	GRD	Grasslands below 1,770 m
White-footed mouse	GORD	
Brush mouse	OFRC	Abundant in forest and chaparral
House mouse	G	Introduced; near buildings
Pygmy mouse	GOR	Grassy habitats to 1,829 m
Northern grasshopper mouse	GRD	Sandy soil
Eastern grasshopper mouse	GORD	Mesquite bajadas, gravelly bajadas

SPECIES	LOCATION	OCCURRENCE
Hispid cotton rat	GRD	High grass cover
Tawny-bellied cotton rat	GR	Lush grasses
Yellow-nosed cotton rat	GOFR	Typically rocky slopes and oaks
White-throated woodrat	GORDC	Needs shrub cover
Mexican woodrat	FC	Nests are inconspicuous
North American porcupine	OR	
Coyote	GOFRDC	
Mexican gray wolf	GOFRDC	Extirpated
Kit fox	G	
Gray fox	OFRC	Typically oak and pinyon-juniper
Grizzly bear	OFRC	Extirpated
Black bear	OFRDC	Usually above 1,675 m
Ringtail	OFRC	Rocky habitats
Coatimundi	OFRC	
Raccoon	GRD	Primarily riparian
American badger	GRD	Typically grassland
Spotted skunk	GOFRDC	Rare
Striped skunk	GOFRDC	Most common skunk
Hooded skunk	GRD	
Hognosed skunk	GOFRDC	Typically canyons
Mountain lion	OFRC	Especially canyons and cliffs
Bobcat	GOFRDC	
Collared peccary (javelina)	GORDC	Typically riparian and woodland at lower elevations
Feral hog	GORDC	Introduced; typically riparian
Mule deer	GOFRDC	Usually lower elevations
White-tailed deer	OFRC	Canyons, rugged terrain, usually above 1,830 m
Pronghorn	GORD	Grasslands

From Moore-Craig 1996: 92–94

Key: Grassland = G; Oak Woodland = O; Coniferous Forest = F; Riparian = R; Desert Scrub = D;
     Chapparral = C

pasture. The site is found on a low bench that runs along the east side of the creek. Immediately west of the creek an almost perpendicular wall of rhyolitic lava forms a protective scarp that shelters the site. Further east of the site lies another high, flat, bench-like plateau that rises to 150 ft above the small valley floor. To the north and south of the site the valley widens into a broad plain, broken occasionally by small hillocks and arroyos.

The site lies on a flattened area approximately 35 m north-south and 18 m east-west. Three 1-m-high mounds occupied the southern periphery of the site and showed indications of partial washing toward Deer Creek. The northern end of the site was badly eroded and, like the southern end, showed a tendency to wash westward toward the creek (by 1999, significant erosion was also found on the northeast edge of the site and was washing east). The surface of the site was littered with sherds, mano and metate fragments, a number of recognizable flakes, and fragmented tools made of basalt, obsidian, and chert. There was some evidence of pot-hunting disturbance especially on the southern end of the site. During the excavation of the Pendleton Ruin by the Cosgroves in 1931, they surveyed the Deer Creek area, located the Joyce Well Ruin, which they referred to as Site #21, "Brushy Creek Ruin" (see Kidder et al. 1949). During this early survey they also found indications of digging at the site.

In 1962 we found two separate historic structures on the southeastern edge of the site: the adobe remains of a homesteader's dwelling and a horse shed. (In 1999, only a portion of one wall of the dwelling remains standing.) According to the local ranchers, this was the original home of a settler named Joyce who came to the Deer Creek area in the 1920s. The well across the creek and west of the site was named after him.

EXCAVATION

During the survey of the site in 1962, we decided, for the following reasons, that the site should be archaeologically investigated the following year. First, the site appeared similar to the Clanton Draw and Box Canyon sites that were under excavation at the time. Second, a greater abundance of identifiable pottery, including a large percentage of Chihuahuan polychrome as well as El Paso and Gila Polychrome, was identified on the site surface than was originally seen at Clanton Draw or Box Canyon. Third, the mounds indicated that a ruin of 40–50 rooms existed with the possibility of more rooms to the south and southeast. Fourth and finally, the site had the overall surface indications of an Animas phase site of importance because of its location and its relation to other sites farther to the south down the Deer Creek drainage, namely the Culberson Ruin.

With the information gained from the survey the previous year, the School of American Research initiated an archaeological investigation of the Joyce Well site in June of 1963. The field party, composed of 10 members, spent three months excavating the ruin as well as surveying other sites to the south near the Deer Creek confluence.

We found the Joyce Well site to be composed of contiguous rooms, varying in depth from two to three rooms, enclosing a plaza. This typical Animas phase layout (see Fig. 1.2) had been found at Clanton Draw and Box Canyon (McCluney 1965a) as well as the Pendleton Ruin (Kidder et al. 1949). Using the techniques refined during our previous excavation, we dug exploratory trenches to expose the walls of the rooms (Fig. 2.1a, b). The first trenches intercepted the walls of the northern tier of rooms. As anticipated, the tops of adobe walls appeared at a depth of approximately 10 cm below surface. At the northern end of the site, however, deflation had reduced the cover so that the tops of the walls were immediately below the modern surface. Soon the first of the rooms was revealed. (We numbered rooms in sequence as excavated.)

FIGURE 2.1. (a) The 1963 excavation, looking southwest, with the historic structure visible in the upper right; (b) The east wing, looking north, of the Joyce Well site, 1963.

Room 1, 3-by-4 m in size, had badly deteriorated walls and floor, but we could still discern significant architectural details. The walls were constructed of coursed adobe slabs, approximately 5–6 cm in thickness, and we estimated that their original height was at least 3.5 m above the floor surface. From the excavation of Room 1 alone, however, we could not determine roofing or roof construction.

Although the floor of Room 1 was badly eroded and had been subjected to prolonged rodent action, we were able to determine that the floor was quite thick, ranging in thickness from 4 cm in the center, to 6

cm in the corners and near the north wall of the room (Fig. 2.2). We left
the floor intact so that it could dry and perhaps disclose additional features obscured initially by moisture.

The floor had been laid on a level bed of sand or fine gravel and then
carefully smoothed and polished, evidence of which was visible on a
small patch of preserved plaster in the southeast corner of the room.
The floor had no features such as fire pits or storage pits. However, we
found two circular depressions, presumably postholes, located near the
north wall. The smaller of the two was plugged with adobe and the
larger was filled with windblown detritus. After clearing the fill from the
larger hole, we discovered that a fine, thick coat of plaster had been applied to the interior of the hole after the floor had been laid down.

Although we assumed that the room lacked the common four-post
system for roof support, a single-post support pattern, with the tops of
the walls of each room bearing the major weight of the roof, became
more apparent as excavation on the site progressed. The smaller
plugged posthole evidently was used to temporarily prevent roof sag,
because the interior of the hole was unlined and appeared to have been
rather hastily constructed.

An appearance of massiveness was everywhere present in this room.
Although deteriorated, it nevertheless presented an impression of sub-
stantiality in design and architecture. The west wall of the room was
very thick, the corners massive, and in several instances buttressing
added additional support to the walls. The room had burned shortly
after abandonment as charred bits of wood and corn lay on the contact
level of the floor. We almost overlooked a hole cut into the south wall,
as it was plugged much like the smaller of the postholes. After removal

FIGURE 2.3. Room 3 of the Joyce Well site.

of the plug, we saw that a ventilator hole, finished with plaster, had been opened to the room sharing the common south wall.

The excavation of Room 1 extended over a period of four days because it served as a "laboratory room" to give the field crew experience in clearing a room that appeared typical of the site. After giving all individuals the opportunity to take part in the clearing of this room, we assigned the remaining rooms of the site to teams of two men.

As excavation progressed and the clearing of the rooms moved southward, more walls displaying additional features appeared. Room 3 was significant among the northern tier of contiguous habitations (Fig. 2.3). Within the room a second enclosure or annex was built. The architectural pattern of this interior room was highly sophisticated for Animas builders. Evidently, an old post had been selected as the corner support for the secondary room. A 50-cm-square column of adobe, built around the post, formed a buttressed corner for the construction of the two walls. The room may have served as a mealing area for the preparation of corn. However, since no metates or depressions for holding the metates were found, an alternative function for this small room could have been a special shrine or place of worship.

By the time the excavation of the site had reached Rooms 16 and 17, the walls had become much thicker and better made. Room 17 was more elongated than any of the other rooms in the site: approximately 2.3 m wide, east to west, and 6.5 meters long, north to south. The exceptionally thick walls of this room were formed of long, thick blocks of puddled adobe. As the room had burned, the wall features and indications of the method of construction were preserved. Thick plaster covered the walls to a level of about 40 cm above the floor in a continuous layer from the center of the floor. The corners, well formed by the process, were sloped at the juncture of floor and wall. (This construction technique is characteristic of the Animas phase pueblos and had been observed previously at Box Canyon [McCluney 1965a].) Evidence of polishing was still clear on the lower areas of the walls near the juncture of the floors (see Fig. 2.4).

The path of our excavations continued to the south, then to the west and north through the rooms adjacent to Plaza 1. We cleared 45 rooms before the end of the season. (Descriptions of the rooms appear later in this report.) In general, however, we continued to find the characteristic architectural features described above.

GENERAL ARCHITECTURAL FEATURES

WALLS

Throughout the Joyce Well site, the occupants built their walls of coursed adobe using a method very similar to what has been described at Pindi Pueblo (Stubbs and Stallings 1953). The adobe units were large, formed on the walls in slabs measuring about 30-by-70 cm, and laid in such a way that absolute joining was accomplished. At a depth of 10 cm below the surface of the floors, on a layer of caliche, we found foundation trenches that had been dug for the walls. A slight belling of the foundation trench lent additional stability to the first course of adobe.

The walls and floors had received careful sustained maintenance, polishing, and finishing: four separate plaster applications had been made on the walls of 32 of the 45 rooms at the site. Again, as at the Clanton Draw and Box Canyon sites (McCluney 1965a) and the Pendleton Ruin (Kidder et al. 1949), we found evidence of palm patting and finger smoothing on the surfaces of many of the Joyce Well walls. The adobe plaster was always of fine texture, and it appeared that great care had been taken to remove small pebbles and other matter prior to the application of the plaster to the walls. In each instance, in rooms that possessed niches, entryways, and other wall features, we observed that the features were cut into the walls after they had been plastered. The evidence for this was that the thresholds and sills had aprons of fine plaster applied to their interiors and brought around to a smooth contact with the wall face. Everywhere the evidence of finished workmanship was apparent.

FLOORS

The floors of the rooms were like the walls, beautifully constructed (see Fig. 2.4). By removing the plaster from the interiors of such floor features as fire pits, storage pits, and postholes, we were able to visualize

FIGURE 2.4. Room 19 showing fire pit, collared post support, and plastered floor.

the processes used in laying the floors, which were identical throughout the site.

First, after the walls had dried and before the application of the finish coat of plaster, a bed of fine sand or gravel was laid on the native surface already smoothed to a uniform level throughout the room. Second, the sand was wetted and packed. We found evidence of several applications of wet sand in the corners to form a cove at the juncture of the wall. Third, after the sand base had dried and set, the first application of plaster was made. This first thick coat measured a minimum of 2.5 cm and was bonded to the second coat by small embedded pebbles. A third finish coat was applied over the first two coats. Its consistency after drying enabled the application of a fine polish to its surface. (In the cases where raised fire pits and collars for posts were constructed, a slightly different technique was used. This method will be discussed later in this section.) While the flooring was being done, the same techniques of plastering were used on the walls to form a continuous, uninterrupted surface (see Fig. 2.4).

We observed reinforcement, by the application of additional plaster, over the areas of the floors that received constant wear (i.e., entryways, the periphery of fire pits, and areas adjacent to the corners). However, this secondary application for strength did not appear to be added on or temporary, as the section to be reinforced had been first excavated down to the first layer of floor plaster and reconstructed back to floor level. We constantly mistook these reinforced areas for intramural burial locations. In all but one case this belief was not borne out.

ROOFS

The question of how roofs were constructed at the Joyce Well site was not completely answered during the excavation procedure. Through our attempt to correlate the placement of the postholes within and among some of the rooms, it became apparent that roofs throughout most of the site were constructed in long spans covering several rooms at one time. This roof plan was evident for Rooms 2 through Room 7 in the northern tier, and for Rooms 12 through 24 in the southeastern area. The alignment of postholes in both cases ran along the north-south axis. These postholes, both collared and uncollared, exhibited diameters of 10.6 cm to 20.4 cm. Such large diameters suggest that the horizontal spans of the roof beams were long, and hence, rafters must have been at least 25 to 30 cm in diameter. As at Box Canyon (McCluney 1965a), we found evidence that large metates had been placed on top of these substantial roofs. Rooms 15, 18, and 19, in contrast, appear to have had individual roofs.

Based on observations of floor contact proveniences and roof fall, it appears that the roofs were constructed in the following sequence:

1. Long beams were placed on top of the walls that spanned, in some cases, three to four rooms simultaneously.
2. Beams were anchored (method not determined because the tops of the walls were eroded).
3. Secondary beams, or stringers (6 cm in diameter), of either juniper, cottonwood or Arizona sycamore were laid over the beams.
4. Green brush and foliage were placed on the secondary beams.
5. Adobe was added and smoothed to a flat surface.
6. Vertical posts were erected within the rooms to limit sag or to furnish support to roofs areas used for activities.

DOORWAYS

Some of the most interesting features of the Joyce Well site were the many door and entryways observed in the rooms. Of the three types of doorways encountered at the site, all were strategically placed either to open into an adjoining room or to provide a direct exit to the plaza in the center of the village. We did not observe a doorway leading directly outside of the village.

The most significant of the three types of doorways were those of the T-shaped variety (Fig. 2.5), a door shape typical of the Animas phase sites excavated in the Animas Valley and reported at the Pendleton Ruin (Kidder et al. 1949), Clanton Draw, and Box Canyon sites (McCluney 1965a). There is little doubt that a southern tradition is evident as this architectural form can be traced from the Casas Grandes area, reported by Di Peso (pers. comm. 1962, 1963, 1964), and the Sierra Madre areas, reported by Lister (1955) in his survey of the sites of the Sierra Madre

FIGURE 2.5. Sealed
T-shaped doorway.

Occidental. This type of doorway is found generally in the Southwest
and especially in the southern Jornada Branch of the Mogollon. We
counted 12 T-doors at the Joyce Well site, all of which were interior pas-
sageways to adjoining rooms.

The second type of doorway in the site was a "half-circle" door cut
into the upper level of the wall, low enough to admit passage but cum-
bersome to negotiate. We call these openings doorways, but there is still
some question about their function. Primarily because of the finished
nature of their construction we placed them in the doorway category, although their function as a ventilator cannot be overlooked.

The third type of doorway had a simple rectangular shape; we found
24 of this type at the site. All, like the T-shaped types, were nicely finished with plaster and reinforced at their sides and thresholds. The average dimensions of the rectangular doors were 60-by-90 cm.

All the doorways encountered were also important because they provided clues to the nature of the village abandonment. We located 13
doors only after the walls of the rooms had been allowed to dry because
the openings were plugged in a way that obscured them during the
clearing of the room fill (see Fig. 2.5). At first no pattern for these
plugged and replastered doors was evident, but toward the end of the
excavation we made an important set of observations. The location of
these plugged doors revealed a pattern that we think is related to the
abandonment of site (see Figure 1.2 and individual room drawings in
the "Room Description" section, below). It appears that the site was
abandoned peacefully, as we recovered no evidence of violence. What is
more, the final abandonment appears to have been made in a series of
departures rather than en masse.

We surmise that, as their fellow inhabitants quit the site, the remaining people banded together for continuing protection by moving into adjoining rooms, plugging the doorways of the abandoned rooms, and completely plastering them into the surface of the walls of the rooms. By observing the filled doorways, we were able to tell whether the doorway had been plastered from the interior of the room or from an adjoining room. It became clear that the last rooms to be abandoned at Joyce Well were Rooms 11, 12, 15, 18, and 21. These rooms were some of the best made and possessed the thickest walls. In every instance, the concern for protection of the remaining inhabitants was evident by the closure of the doorways of the empty rooms.

Collared Posts

Collared posts, like the T-shaped doorways, are typical of the Animas phase villages of southwestern New Mexico and were also observed at the Box Canyon site (McCluney 1965a). The excavated collared posts were carefully executed to serve as additional foundation for the upright roof support posts (Fig. 2.6). The collared posts at Joyce Well, as well as at Box Canyon (see McCluney 1965a for details of construction), were formed at the same time as the plastering of the walls and floors.

Unlike other Animas phase settlements, we encountered two collared post styles. The first style was almost square (see Fig. 2.4) in outline and measured, on average, about 30 cm on a side. The second style was circular in shape and was found with smaller posts (Fig. 2.6). We uncovered nine circular collars and they averaged between 11 and 12 cm in diameter. They were confined generally to the smaller rooms of the site.

Raised Fire Pits

We found three raised fire pits, which is another feature common to Animas sites. Room 18 had not only the best example of a raised fire pit, but in this case this feature was in line with both a collared post and a T-shaped door (Fig. 2.6). This raised fire pit was remarkable in both execution and detail. Located in the center of the room, the raised portion had a square outline and extended 3 cm above the floor. A slight flaring toward the eastern end made its overall measurements 53-by-59 cm. The pit was 20 cm in diameter and 10 cm in depth. Entering the pit on the east side of the raised area was a well-plastered ash ramp decorated on its sides by a scalloped design cut into the tapered side walls. Each surface and side of the feature had been replastered several times with great care. The surface of the raised portion had been smoothed, the edges slightly rounded, and the plaster brought gracefully to the floor. The area of the floor surrounding the fire pit had also been raised slightly by several additional applications of plaster. Whether or not this type of fire pit had a special religious or ceremonial significance among

FIGURE 2.6.
(a) Raised fire pit
and collared post
in Room 18;
(b) close-up of fire
pit.

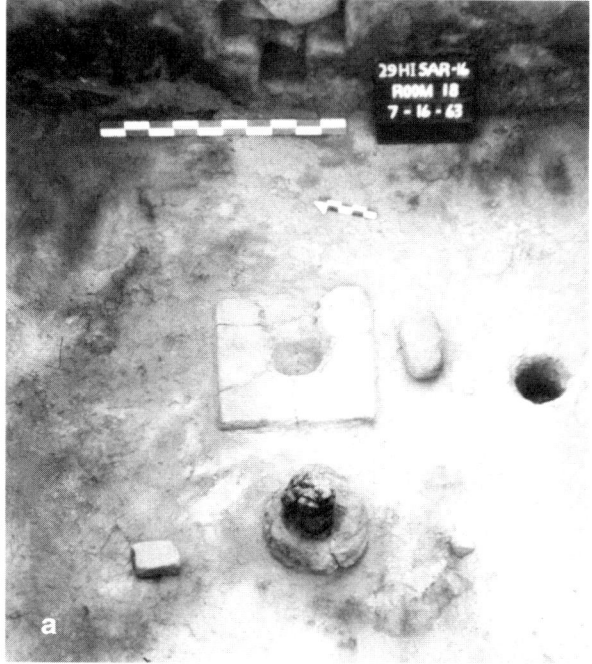

the Animas people is not known. A similar raised fire pit was encountered in the Box Canyon site (McCluney 1965a) but lacked the embellishments of the fire pit of Room 18 at Joyce Well.

The other type of fire pit excavated in the site was the simple circular
feature dug directly into the floor surface and plastered inside. Regardless of the type of fire pit observed, each was formed with utmost care
and was well plastered.

It must be clearly stated here that, based upon the foregoing descriptions of the floors, walls, doorways, and fire pits of the Joyce Well site, the Animas people were advanced in the techniques of village layout and especially in the architectural refinements of finished plastering and construction of floor and wall features within the rooms of their villages.

Central Plaza

The consistent village pattern for the Animas Phase consists of a series of contiguous rooms constructed around a central plaza with one side of the plaza either open or reduced to a restricted passage at one or both sides (see Fig. 1.2). Examples of this typical layout were noted at both the Pendleton Ruin (Kidder et al. 1949) and the Box Canyon site (McCluney 1965a). The plaza appears to have functioned in a number of ways. First, the architectural arrangement limited direct passage to the inside of the village compound, which is a more defensible arrangement than a linear room block. Second, everyday activities of the village were likely carried out in the plaza where the occupants could congregate in an area sheltered from the elements (oppressive wind) and protected from unwanted outsiders. It was noted at Box Canyon (McCluney 1965a) that the hearths were used primarily for heat and light, which suggests that cooking could have been done in the plazas. Even in the winter, the arrangement of the rooms around the plaza provides ample protection from the cold. Third and finally, it is possible that religious ceremonies and related activities, such as dancing, games, and other observances, were performed within the confines of the plaza.

As a result of the short field season, little excavation was done in the plaza. We did, however, observe flagstones in Plaza 1 near Rooms 13 and 39 while clearing the exterior of the walls. At the Pendleton Ruin (Kidder et al. 1949) this same type of flagstone floor existed in the plaza. We found only one possible doorway into the Joyce Well plazas. We must conclude that the occupants gained access to the plaza by way of roof hole and ladder.

ARTIFACTS

Chipped Stone

Approximately 2,800 items of chipped stone (projectile points, scrapers, and choppers) were recovered from Joyce Well. Most of these artifacts, especially the projectile points, were found within the fill of rooms. It appears from this occurrence that most of the serviceable projectile points were carried away by the last of the inhabitants during their exit from the site. Since some of the chipped stone artifacts are under question as to whether they belong to the Animas phase typology, they will be discussed according to their appearance rather than on the basis of nomenclature comparable with other artifacts recovered from other Animas phase sites.

Projectile Points

The projectile points from the Joyce Well site were typical of the generalized Mogollon types recognized by Wheat (1955) early in the Mogollon area. Rarely did we find points with distinctive corner notches producing stemmed haft elements. Most of the points were small and triangular, suggesting use on small game. Of the 25 points collected from the site, 20 were manufactured from black obsidian and 5 from chert or jasper. Although these points came primarily from room fill, three were found at floor contact and several more were found in caches or wall niches.

Several large deer bones found in firepits within rooms indicated that large game animals were hunted or trapped by the Animas people. Large amounts of rabbit bone, however, were found throughout the site (especially in firepits) suggesting that smaller game of this type were more often hunted. The average length of the points described is 2.6 cm, which is a size better suited for rabbits and small game.

Knife

One knife is part of the collection and it was recovered from the floor of Room 23. The point is 6.6 cm long (tip missing), 2.9 cm wide, and 0.05 cm thick, and it is long and ovate in cross section. It was produced from a dark-pink rhyolite and its shape is elongate-triangular and the base is slightly convex.

Scrapers

Of the 44 scrapers found in the rooms, 18 were recovered at floor contact and the remaining 26 from the surface and room fills. The sizes of these scrapers ranged from 8.4 cm to 3.4 cm in diameter. Most of the scrapers were made from large- to medium-sized flakes, but the smallest were flaked from fragments of other tools. None were combination scrapers denoting multiple use. The most common material was chert, but two of the small scrapers were made from obsidian. Most of the scrapers were well chipped and formed for repeated use.

Choppers

Large choppers occurred with frequency at the Joyce Well site and we found seven at floor contact. Usually fashioned on cores, these tools were probably utilized in a number of ways including chopping of wood

and preparation of food. Three showed utilization as large hammerstones after having been discarded as choppers. Why there should appear such large tools as the choppers was baffling when we considered
the almost complete absence of large scrapers and projectile points in
the site.

Cores

Of the 36 cores recovered from the site, the majority occurred in room
fill. No evidence of utilization as planing or pounding tools was evident.
All were composed of hard material such as chert or quartzite. Evidently, these cores were the residuals from the manufacture of the points
and scrapers.

GROUND, PECKED, AND POLISHED STONE

Manos

Manos were frequently recovered both in room fill and at floor contact
throughout all of the rooms that displayed occupation to any degree. A
total of 78 manos, either fragmentary or whole, were recovered and we
classified them into two major types.

Fifty-one manos were of the ovate-elongated variety (Type 1), which
was described in the Box Canyon report (McCluney 1965a). They had a
mean length of 10.8 cm and mean thickness of 1 cm. Their ends and
sides were pecked into a graceful contour, and on most specimens the
ends showed extensive wear. Because most of the manos were made of
vesicular basalt and were very light in weight, we concluded that they
were of the one-hand variety. Six of the specimens were made from a
fine-grained sandstone.

The remaining 27 manos we refer to as Type 2, which were oval and
unifacial. Most of the manos of this type were made either of fine-
grained sandstone or equally fine-grained rhyolite. Only a single specimen of this type was made of basalt. Two distinct methods of grinding
were evident. The first used a push-pull motion, as indicated by faceting
along the edges. The second method utilized a rotary motion, which was
inferred from wear on one or both faces.

Metates

The 39 whole or fragmentary metates recovered at the Joyce Well site
were of two distinct types. The first type, of which we found 23, was the
large, one-end-closed variety with deep troughs and sloping sides. The
materials most commonly used in their manufacture were vesicular
basalt and granitic or granite-like material akin to a heavy rhyolite. The
majority were made of the black basalt, which occurs within a few kilo-

FIGURE 2.7. Polishing stones.

meters of the site. We found many on the floors of the rooms; but four were located in room fill, four appeared in the plugged doors, and one served as a covering slab for a connecting doorway between rooms. Most showed signs of heavy usage, and two were extremely thin.

The second type of metate was a single slab, which appeared more like a flattened, unfashioned piece of sandstone than a finished metate. Upon examination of these specimens, however, we discovered that each contained a shallow working basin. Sixteen of this type occurred at floor contact and in the fill of six rooms.

We surmised from the occurrence of the two types of metates that the occupants of the Joyce Well site devoted a large amount of their time to the process of grinding cereal grains. The large number of metates and manos found, considering the population of the village, indicated concentrated activity in the preparation of foods by grinding and pounding.

Pounding Tools

Of the recovered artifacts used for pounding grains and other foods, all were well-worked specimens made from sandstone or rhyolite. Their ends showed evidence of use in pounding or pestle-like motion. Whether or not they were used in mortars is questionable as we found no mortars in the site. It is conceivable that bedrock mortars used with the pounding implements were located somewhere away from the site.

Polishing Stones

Nine polishing stones fashioned from small cobbles of rhyolite and chert made up an interesting collection of diagnostic artifacts from the site. We found them mostly at floor contact in Rooms 20, 24, and 27 (Fig. 2.7). Significant as a concentrated assemblage, this collection might indicate a task-specific area within the site for smoothing and finishing of pottery and other artifacts of clay. One specimen, in particular, was

FIGURE 2.8.
Hammerstones from
the Joyce Well site.

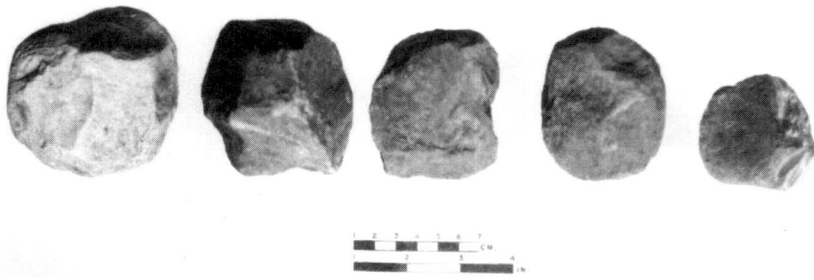

notable. It was made of black rhyolite and had three faceted sides. Under microscopic examination, rotary striation marks were evident. The tapered end of this tool was slightly battered, showing its additional use as a pressing or indenting tool. All of the polishing stones were well made and all demonstrated long use.

Hammerstones

The hammerstones were of a simple variety made from selected cobbles of rhyolite and chert (Fig. 2.8). We found them in abundance in all levels of room and floor fill as well as at floor contact. Most of the whole specimens showed evidence of prolonged use, as batter marks had obliterated the ridges and peaks of the original fragmentary cobble.

FIGURE 2.9. Shaft smoother.

Shaft Smoothers, Awl Sharpeners, and Whetsones

Of the four shaft-smoothing stones recovered during excavation, we found three on floors of rooms (Fig. 2.9). The largest of the smoothers was manufactured from a dark gray rhyolite material that had been previously utilized as a mano. A distinctive groove, which distinguishes the shaft smoothers, was deeply worn down the long axis of one face. The three other smoothers were made of sandstone or volcanic tuff. Accompanying the largest shaft smoother was an awl sharpener. It was made of dark red sandstone and it was well formed on all sides and had rounded ends. Deep, narrow grooves ran down three sides of the tool. A second, badly fragmented awl sharpener was found in room fill.

Two whetsones, one of volcanic tuff and the other of black basalt appeared in Room 32. Both artifacts had been carefully shaped into a rectangular form. Almost obscure facets indicated their use as tools in honing and smoothing.

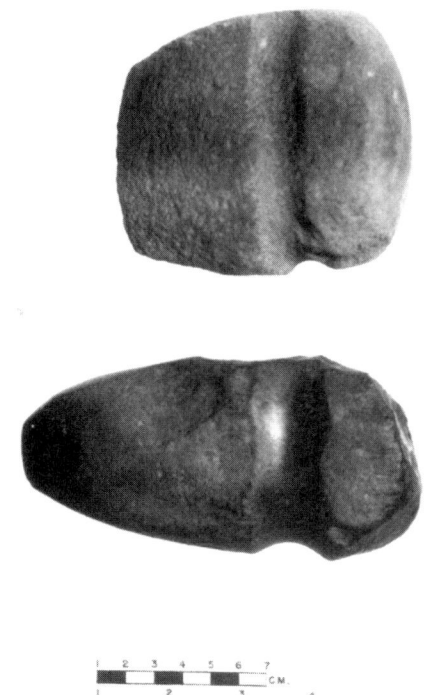

FIGURE 2.10. Three-quarter grooved axe of dark-green rhyolite (bottom) and a stone maul (top).

Axe and Maul

We found one axe, three-quarter groove variety, in the fill of Room 14 (Fig. 2.10). The axe, made of dark-green rhyolite, showed signs of prolonged and heavy use and it had been badly fractured on one face and on one corner of its poll. Polish on the groove suggests that the axe had been hafted. Although in overall appearance the axe was well made, its poll was crudely shaped. The long flake scar on one face suggests that the axe had been damaged by a direct shock-contact blow but then resharpened for additional use.

We recovered what appeared to be a maul from the floor of Room 22 (Fig. 2.10). Although it was shaped like the axe described above, it was not made of hard stone, but rather from medium-weight volcanic tuff. We noted with interest that the edge, rather than being blunt like most mauls, was ground to a rather sharp edge. The tool had a three-quarter groove, and a nicely shaped poll, but it is unlikely that the axe was used in timber-felling operations. We think that the material would not withstand such pounding and that it is more likely that the tool was used for something like pounding or beaming skins.

Palettes

Five paint palettes were recovered from various locations. One came from room fill, three were on the surface, and one at floor contact in Room 7. All were of the plain variety with no embellishments of design or decorative incising (Fig. 2.11). The palettes were fashioned from slabs of sandstone and were likely used for grinding of natural ochres and other materials used for paint. Two palettes had red and yellow stains on both faces.

FIGURE 2.11. Palettes.

Undecorated paint palettes are common in the Animas area, which contrasts with the more highly decorated palettes in the Mimbres and Hohokam region. All of the samples from Joyce Well appeared to be fragmentary but their edges and corners were smoothed.

Pendants

Three pendants of stone were recovered from room fill. One pendant of note was shaped in the form of a bird (in profile) with incised designs on the head and sides of the body. A great deal of effort went into forming and polishing the stone into this form. Its uniqueness was enhanced after it was determined that it is made of serpentine, which is not found locally. The nearest source of serpentine is central Mexico. A biconically drilled hole was deformed through use-wear into a teardrop outline, suggesting that the pendant was worn for a long time.

The second pendant, made of sandstone, was in the form of perhaps a dog or bear. Again, it was quite reminiscent of pendants occurring farther to the north. A large biconically drilled hole, located near the upper center of the back of the animal form, showed signs of prolonged wear. In comparison to the bird pendant, however, this one had a more crude appearance and was more roughly made. The third pendant, recovered from Room 15, was in the shape of a sphere with a small knob on top. Use-wear traces on the constricted juncture of the two spherical forms suggests that the pendant was suspended by wrapping a cord around the juncture. The pendant was made from light blue-green fluorite, which is not found locally, and was expertly finished.

Miscellaneous Objects

We found eleven objects of stone that had to be assigned to the "problematic" category in terms of identification. These items, which will be only briefly discussed here, were made of materials that ranged from chalcedony to sandstone. Included in the inventory were miniature ball-shaped objects, small triangular bits of rhyolite, three unfinished pieces of pyrite, and several disc-shaped fragments of pitchstone. Since all of these artifacts were highly fragmentary, identification was difficult. These artifacts do, however, speak to the artistic skill of the Animas people both in design and sculpture. We assumed that these items were used in religious ceremony or some other special activity.

BONE

Awls

An additional category of artifact that gave important information about the behavior of Joyce Well villagers were the awls manufactured

FIGURE 2.12. Bone awls from the Joyce Well site.

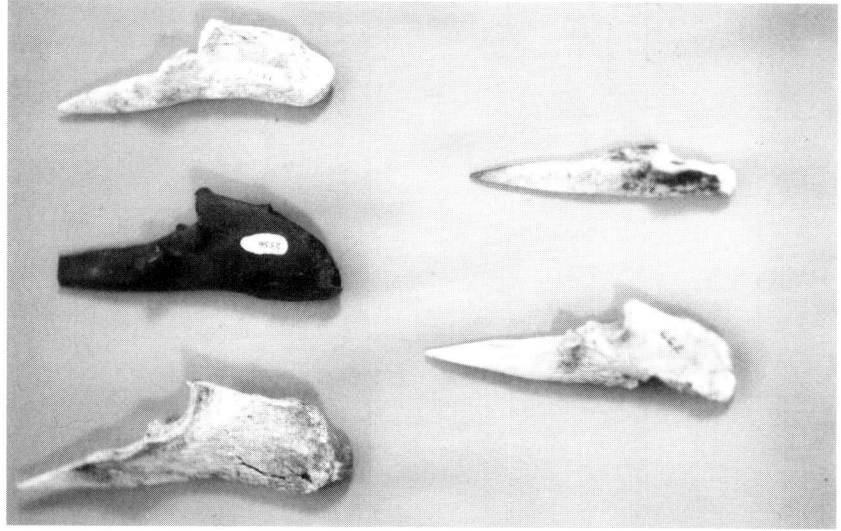

of bone (Fig. 2.12). We recovered 21 specimens during excavation, and we placed the awls into three main types.

Type 1 awls were manufactured from deer ulnae and were unmodified except for pointing of the shaft. All but one of the five from this type were whole and intact. This type of awl, recognized by Wheat (1955), occurs early in Mogollon culture and has been reported from the Mimbres area, north of Joyce Well. The specimens we recovered appeared to be well used because the shafts had a high gloss.

We found four awls of Type 2, one of which was highly fragmented as a result of burning. Probably manufactured from the proximal end of a deer metapodial, the Type 2 awls had heads that appeared to have been only partially modified. These awls were well made, and they also had a glossy polish likely caused by friction with skin or hide during sustained use.

Manufactured from split bone, probably deer, the Type 3 awls were relatively short compared to the other two types. Fracturing of the bone produced an edge that might have been used as a cutting or trimming knife. We recovered three awls of Type 3 awls from room fill.

Notched Bone

One specimen of notched bone, apparently made from a discarded awl, occurred in Room 16. It had ten horizontal incisions across the face of the shank, perhaps made by sawing. We inferred that this artifact may have been used as a musical rasp. As definitive information is lacking from the Animas area regarding notched or cut bone, its function as a reaming tool or a specialized weaving tool is possible as well.

FIGURE 2.13.
Necklace of slate and
hematite beads with
turquoise dividers
(on left).

Turquoise, Slate, Hematite, and Shell

Almost all of the specimens within this category were grave furniture and adornment found in excavating intramural burials. Only occasionally did we find pendants, beads, and fragmented shell in room fill or at floor contact.

The most outstanding objects manufactured of turquoise were pendants, beads, and polished fragmentary slabs. Turquoise beads in particular appeared in combination with beads of other material used in the assembly of necklaces and wristlets. One particularly fine necklace of a single strand, recovered from a burial, demonstrated the pleasing effect achieved by the combination of black and red beads with larger turquoise beads serving as occasional dividers (Fig. 2.13). Although turquoise is reported for Hidalgo County (Northrop 1959), there is no identified source adjacent to Joyce Well nor in the Deer Creek area. The Animas people may have journeyed northward for their supply of this material into the Hachita and Burro Mountain areas.

A small wristlet recovered from an infant burial was composed of turquoise beads of a variety of shapes, including tabular, tear-drop, tapered cone, circular, and cylindrical (Fig. 2.14). Whether the turquoise material was brought to the site and finished into pendants and beads, or finished products were traded for in the areas to the north, is not known. The almost total absence of waste turquoise within the site, however, points toward the acquisition of formed pieces. For adornment, turquoise objects tended to be less popular than slate and shell, either because of local preferences or because turquoise was much more difficult to obtain.

We found two pendants of jadeite near the southeast corner of Room 21. At first, we assumed that the pendants were made from turquoise, but microscopic analysis and specific density measurements identified the items as jadeite. Jadeite is common in northern Mexico, which was the likely source of the Joyce Well specimens.

Slate and hematite beads occurred in three burials. These small, well-worked beads were used in the assembly of necklaces either in the two-color combinations of red hematite and black slate, or in combination with the turquoise divider beads described above. Measuring less than 3 mm in diameter, these beads were nevertheless drilled biconically and strung with sinew or some other material. We found a small wristlet of slate and hematite arranged with *Olivella* shell beads in the burial of a female adult (Fig. 2.14).

The source for the hematite and slate is undetermined, but it is

FIGURE 2.14.
Necklaces and
wristlets of shell and
turquoise, inferred by
McCluney and his
team.

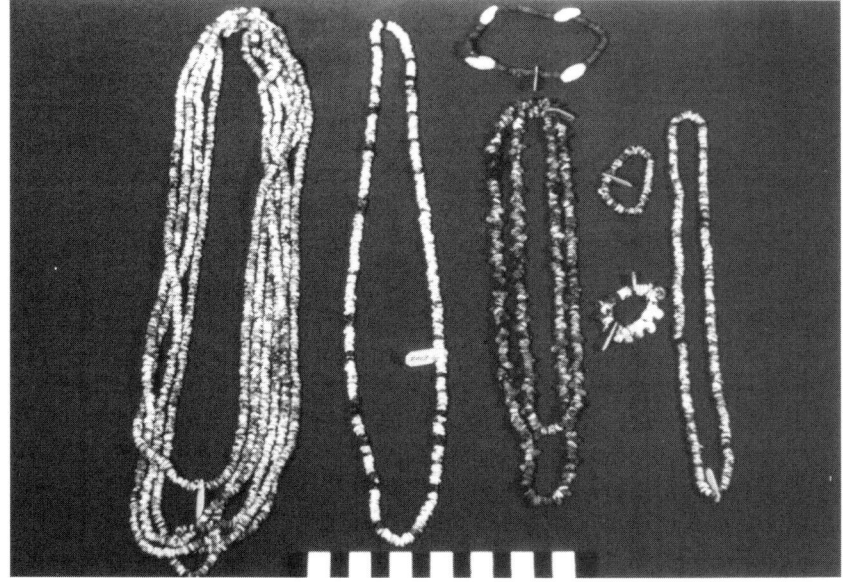

very likely that these beads were traded from some other area. We
found no indications of workshops at the Joyce Well site for any of the
jewelry.

Shell beads made from *Glycymeris*, *Nassarius*, *Olivella*, and *Ostrea*
occurred in the burials, in scattered units, and were assembled into
necklaces and wristlets (Fig. 2.14). Shell was the most common material
used in all ornamentation.

Of the five samples, which included one wristlet and five necklaces,
we found two in burials. The shell necklaces of *Glycymeris* and beads
ranged in complexity from single strands, to double strands, and one
with five strands. One unusual necklace included *Glycymeris* and *Nassarius* beads in combination. A wristlet, accompanying an infant burial,
was made entirely of *Glycymeris* beads.

POTTERY

As a result of three months of intensive excavation, we recovered a total
of 9,985 sherds as well as 21 whole or restorable vessels. Tables 2.2 and
2.3 list the 40 types of pottery found and the numbers of sherds and
whole vessels in each type.

The first excavation of an Animas phase pueblo in the Bootheel of
New Mexico (Kidder et al. 1949) identified locally made Chihuahuan
polychromes. This led them to conclude that Pendleton Ruin was an
outpost of the Casas Grandes culture of northern Mexico, or that the
occupants of the ruin were local traders who gained Chihuahuan pottery through trade. The excavations at Clanton Draw and Box Canyon

TABLE 2.2. Ceramic Types and Counts

TYPE	NUMBER OF SHERDS
Alma Plain	4
Babicora Polychrome	8
Casas Grandes Corrugated	34
Casas Grandes Incised	195
Casas Grandes Incised-Corrugated	72
Casas Grandes Obliterated-Corrugated	170
Casas Grandes Plain	63
Casas Grandes Scored	7
Casas Grandes Tool-Punched	47
Chupadero Black-on-white	18
Circle-Tooled plain ware	3
Cloverdale Corrugated	118
Corralitos Polychrome	3
Cord-Impressed plain ware	2
Cord-Marked plain ware	9
Dublan Polychrome	?
El Paso Polychrome	939
Gila Polychrome	352
Gila Red Ware	13
Huērigos Polychrome	4
Imitation Gila Polychrome	27
Mimbres Corrugated	1(?)
Playas Red	934
Playas Red (Brushed Exterior)	1
Playas Red (Incised)	113
Ramos Black	160
Ramos Polychrome	1483
St. Johns Polychrome	6
Tucson Polychrome	4
Unclassified brown ware	5062
Unclassified brown ware (polished)	1
Unclassified corrugated	1
Unclassified incised	1
Unclassified neck-banded	1
Unclassified plain ware (brushed interior)	12
Unclassified paddle-and-anvil	20
Unclassified smudged exterior	2
Villa Ahumada Polychrome	8

(McCluney 1965a) found similar types of ceramic assemblages. The Joyce Well ceramics add to this basic database.

The pottery types from Joyce Well were classified into two major groups: culinary and decorated. Each group will be discussed as a whole with additional comments for particular pottery types.

TABLE 2.3. Whole Vessel Counts

CERAMIC TYPE	NUMBER
Carretas Polychrome	1
Cloverdale Corrugated	5
Cord-Marked plain ware	1
El Paso Polychrome	2
Gila Polychrome	2
Playas Red Incised	2
Ramos Black	2
Ramos Polychrome	5
Tucson Polychrome	1

Culinary Pottery

Culinary pottery comprised the largest percentage of the ceramic inventory. The dominant type was brown ware. This plain, untextured ware was made from a local clay and contained temper composed mostly of feldspar and small amounts of basalt. In the fired vessels, there was a slight tendency for the surface to crack or craze. During manufacture of the brown ware, the surface was brought to a high polish that was retained after firing. Although we anticipated a preponderance of Alma Plain at Joyce Well, Jornada Brown, which is found in the Tularosa Basin far to the east of the Animas area, was the most dominant type. There are, however, some subtle differences in the Joyce Well-made Jornada, most notably in the type of temper as a result of local manufacture. Otherwise, the simple bowl, jar, and small olla forms were similar to the traditional styles of Jornada ceramics. We found sherds of this ware at all levels of excavation and generally scattered on the surface of the site.

The next most abundant culinary ware found throughout the excavations was Cloverdale Corrugated (Kidder et al. 1949). As was noted at Pendleton Ruin and the Clanton Draw and Box Canyon sites, this pottery appears to have been locally designed and made. In form, the samples we found were the simple deep bowl, seed jar, and shallow bowl (Fig. 2.15). A single seed jar of the Cloverdale Corrugated type, found as grave furniture, contained a gourd dipper and cornmeal. Rather than repeating the nomenclature of this type, we refer the reader to the descriptions found in the Pendleton Ruin report (Kidder et al. 1949). Note that the Cloverdale Corrugated pottery of the Animas area lasted throughout the early, middle, and late development periods of the Animas people.

Next in abundance were the six types of culinary wares derived from Casas Grandes: Casas Grandes Obliterated-Corrugated, Incised, Corrugated, Scored, Tool-Punched, and Plain. Types named here are those recognized by Di Peso from the excavation of the Casas Grandes site.

Figure 2.15.
Cloverdale
Corrugated vessel.

The high frequency of Casas Grandes culinary wares at Joyce Well point to several important observations. Although the Casas Grandes culinary pottery in the Animas area was locally made and contained the same local tempers as the other types of Joyce Well culinary ware, the traditional texturing of the local ware is almost identical to the Casas Grandes-made pottery (Di Peso 1974; Di Peso, pers. comm. 1964). Most of the traditional Casas Grandes vessel shapes were present at Joyce Well, the exceptions being the large ollas and the highly erratic forms present at the Casas Grandes. Hence, we infer that the Animas people were part of the Casas Grandes interaction sphere.

Forming a third group of culinary wares were those that do not belong to types originating in either the Casas Grandes or Animas areas. The major distinguishing characteristics of this group are the texturing process used and the variations of texturing and scoring used in the decoration. The types assigned to this category are cordmarked plain ware, and a number of unclassified wares that we list using only descriptive terms: brown ware, paddle-and-anvil, plain ware with brushed exterior, brown ware (polished), neckbanded, corrugated, incised, and smudged interior. As is apparent, we could not identify these types based on comparisons to other regions. The designations for these pottery classes are not type names, but simply descriptive titles used for continuity of this report. At present this pottery should be considered locally made.

The fourth category of culinary ware is small in number but it includes those that had been recognized elsewhere. These are Alma Plain and Tool-Punched Plain Ware. Alma Plain was recognized by Wheat (1955) as making an early entry into the Mogollon area of the Southwest and lasting until Mogollon 3, when this pottery type decreased in frequency. The Alma Plain sherds recovered from Joyce Well, however, indicated that this pottery was used by the occupants of the village. Tool-Punched Plain Ware was made by impressing with a hollow bone or piece of wood or cane to form a pleasing design of scattered circular impressions, apparently confined to the neck of the vessel. This ware, which has also been reported from Casas Grandes (Di Peso et al. 1974), was represented at Joyce Well by four scattered sherds from room fill.

Decorated Pottery

Of 9,985 sherds recovered from excavation, we assigned 4,097 to decorated types. The decorated pottery was the single most important artifact category for understanding the place of the Animas phase in the framework of the heretofore little-known archaeological area. Although the radiocarbon dates (and other dates) that have been derived from materials recovered from Joyce Well are important for defining the chronology of the site, cross-dating the decorated pottery from the site serves that purpose in addition to providing important clues to the behavior and interaction of the prehistoric inhabitants. The salient information gained from a complete study of the whole inventory of decorated pottery will be presented here after discussion of the most significant decorated types. The following discussion is organized based on the presumed point of manufacture.

Locally Made Pottery

After a comprehensive analysis of the decorated pottery, including examination of the temper, paste, and decoration styles, we recognized four kinds of local decorated pottery: Ramos Polychrome, Ramos Black, Playas Red, and Playas Red Incised.

Ramos Polychrome. This ware was first recognized as Chihuahua Polychrome by Brand (1935) and later named Ramos Polychrome by Sayles (1936b). It has been generally accepted that this type was confined to northern Mexico, the Casas Grandes site, and several smaller sites. Some of the diagnostic characteristics of this pottery include the exceptional brushwork in design application, the distinctive three-color design, the wide range of vessel forms, and its extensive distribution. At Joyce Well we found 1,483 Ramos Polychrome sherds and five whole or restorable vessels (Figs. 2.16 and 2.17). It was the most common of the decorated pottery types recovered from the site, and the ceramic assemblage that provided the most important analytic information.

During the initial excavation of the site, when Ramos Polychrome began to emerge, we thought that its presence was a representation of trade relationships with either the main Casas Grandes site or other Casas Grandes villages of smaller size in northern Mexico. The excavators of Pendleton Ruin (Kidder et al. 1949) and Box Canyon and Clanton Draw (McCluney 1965a) had concluded that Ramos Polychrome

FIGURE 2.16. Ramos
Polychrome bowl.

FIGURE 2.17. Ramos
Polychrome jar.

pottery was an indicator of direct contact with Casas Grandes. This perception changed, however, as a result of laboratory analysis and a more careful assessment of the Ramos Polychrome ceramics from Joyce Well. We concluded that a substantial amount of the Ramos Polychrome from the site was locally made. Only 300 of the almost 1,500 Ramos Polychrome sherds were not made locally.

Two attributes were most important for determining manufacturing location of Ramos Polychrome: temper and vessel form. Temper type was certainly the more diagnostic attribute. The author made a comparison between sherds found at Joyce Well and sherds from the Casas

Grandes site furnished by the Amerind Foundation in 1963. Ramos Polychrome from the Animas area had temper minerals assignable to the Quaternary depositions of quartzose porphyry recognized by Schwennessen (1918). In combination with this quartz-like temper were medium amounts of basalt with stains of reddish brown or blue-gray inclusions. So far as is currently known by the author, this distinctive basalt is confined to the Animas Valley and has not been recognized in other adjacent regions. The Casas Grandes sherds, and others collected from northern Chihuahua by the author, in contrast, have medium-size temper particles of quartz, traces of mica, and flecks of reddish-colored hematite or another iron-bearing rock. Thin-section petrographic analysis of the temper was not required because inspection with a hand lens was sufficient for identification.

In terms of vessel form, all of the whole or restorable Ramos Polychrome vessels we recovered were either simple bowls with slightly incurving rims or small ollas. The absence of large storage ollas of either Ramos Polychrome or any other Casas Grandes associated pottery is significant. It is the opinion of the author that the existence of commercialized trade of any nature can be recognized by the presence of large ollas. At Joyce Well we identified no large Ramos Polychrome ollas.

Ramos Black. This pottery type was well represented at Joyce Well and, like Ramos Polychrome, this was at first assumed to indicate trade with Casas Grandes or related sites in northern Mexico (Fig. 2.18). Upon closer analysis in the laboratory we discovered that the identical temper used in the locally made Ramos Polychrome occurred in the sherds and the two whole vessels of Ramos Black from Joyce Well. We noted typical high luster and simple bowl shapes, but the temper difference was clear, and, consequently, highly significant.

FIGURE 2.18. Ramos Polished Black jar.

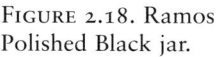

FIGURE 2.19. El Paso Polychrome bowl.

Playas Red and Red Incised. The third locally made decorated ware is Playas Red (including Playas Red Incised), which contributed a total of 1,047 sherds to the pottery inventory. The temper described above for Ramos Polychrome and Ramos Black appeared again in the Playas Red and Playas Red Incised. Playas Red and Playas Red Incised not only have significantly different designs but different vessel forms as well. The dominant form for Playas Red, based only on sherds, was small, simple bowls, whereas the dominant form for Playas Red Incised was large ollas.

Trade Ware

El Paso Polychrome. This pottery type, recognized and named by W. S. Stallings (1931), was also found in relatively high amounts at Pendleton Ruin (Kidder et al. 1949) and Clanton Draw and Box Canyon (McCluney 1965a), the three Animas phase sites on the west side of the Animas Mountains. The occurrence of four large, highly fragmented El Paso Polychrome ollas in Room 31, and two bowls found in room fill suggest trade with people to the east of the Joyce Well site (Fig. 2.19). Analysis indicated that the El Paso Polychrome ollas were not made by the inhabitants of the site, but were present by virtue of trade. It is likely that most of the ollas were used for the transport of dry, rather than liquid items, as they have porous and water-permeable vessel walls. The size of the ollas would also facilitate easy transport, despite their fragility. Given the abundance of corn in both the El Paso and Animas areas, we think it is possible that corn was the commodity of exchange between the two areas. In comparing the sherds of El Paso Polychrome with those from the southern Tularosa Basin in New Mexico, we found the pottery from Joyce Well to be much better made. The walls of the Joyce Well ollas were extremely thin (2 mm) in relation to the overall

FIGURE 2.20. Tucson
Polychrome (b) and
Salado Polychrome
vessels (a, c) from
Joyce Well.

size of the vessels (40.5 cm mean diameter). Further, their decoration was much better than on the sherds from the areas of the type site.

A number of questions arise from the appearance at Joyce Well of the El Paso Polychrome. Did the El Paso people trade with the Animas people in the sphere of "dependence through cooperation"? Were the El Paso people forced peacefully or otherwise to pay tribute? If so, was food, possibly in the form of corn, specified as the tribute item? Finally, and most important, were the El Paso people, like the Animas inhabitants, contributing to the maintenance of the people of the main Casas Grandes site in Mexico? These questions are not likely to receive definitive answers for some time. However, they arise logically and will be addressed in the summary.

Chihuahuan Polychromes. A number of Chihuahuan polychromes were also found at Joyce Well and likely came to the site via exchange from the south. The types were: Babicora Polychrome, Carretas Polychrome, Corralitos Polychrome, Dublan Polychrome, Huĕrigos Polychrome, Madera Black-on-red, Medanos Red-on-brown, and Villa Ahumada Polychrome (Sayles 1936b).

Gila Polychrome. At Joyce Well, this pottery type appears to be non-locally made. Its appearance pointed to relationships between the Joyce Well people and the inhabitants of the Gila Basin far to the west and northwest of the Animas and Hachita valleys. We found one small Gila Polychrome bowl in room fill and a large, fine Gila Polychrome jar at floor contact in Room 19 (Fig. 2.20a, c). Our analysis suggested that these vessels were not locally made. Other researchers recognized locally made variants of Gila Polychrome, but this does not seem to be the case here. Di Peso (pers. comm. 1963–64) recognizes the variant form from Casas Grandes. However, the vessels and the great majority of Gila Polychrome sherds from Joyce Well were of the "formal" Gila Polychrome type (Gladwin and Gladwin 1930).

Tucson Polychrome. This ware was represented by a few sherds and one whole vessel associated with Burial 1 (Fig. 2.20b). Again, as with Gila Polychrome, the presence of this ware at Joyce Well denotes a relationship, probably in trade, between the Animas people and the people of the San Pedro Valley.

Conclusions

Four salient points arose from our study of the pottery of the Joyce Well site.

1. The pottery as a whole represents, in equal proportions, both the local manufacture of pottery and the exchange of pottery from other regions.

2. The appearance of Ramos Polychrome as a locally manufactured type is significant, and sheds light on the relationship between Joyce Well and Casas Grandes.

3. The presence of El Paso Polychrome indicates that trade with the region to the east of the site was important and that both the El Paso and the Animas peoples were mutually dependent, through trade, on the Casas Grandes people to the south.

4. The pottery assemblage as a whole suggests that occupants of the site were either Casas Grandes people or their close relatives inhabiting a northern peripheral village.

BURIALS

Twenty-three burials made up the inventory of skeletal material recovered from the Joyce Well site. An additional three burials were observed but could not be removed because of the fragmentary nature of the bone. Although the skeletons were interred intramurally, the poor condition of the bone was a constant barrier to removal and preservation. We cleared all of the skeletons by degrees so that rapid drying would not deter the process of sketching, measuring, and photographing.

In general terms, the burials were located under the floors of rooms. There was neither systematic placement of the individual with regard to orientation of the body, nor was there attention given to the burial position. For example, we found one skeleton in a sitting position, another with arms over chest in a semi-sitting position, and a third fully extended on its back. The majority were placed in well-formed burial cists that were boot-shaped in profile. Semi- or fully extended postures were also observed and they were seen in elongated cists. Grave furniture consisted primarily of pottery vessels and ornamentation, generally in the form of necklaces and wristlets.

There was no suggestion of violence as the cause of death in any of the burials. Natural articulation was the general rule. In three of the burials we found no grave furniture or other forms of offering. The reason for the lack of grave goods could not be determined. A more complete discussion of the recovered human remains can be found in Chapter 4.

PLANT REMAINS

From the initial clearing of the northern rooms of the site, corn began to appear as both scattered kernels and fragmentary cobs, some with kernels intact. All of the corn recovered from the site had been charred as a result of burning. From our assessment of the number of grinding tools, we concluded that the growing, cultivation, and mealing of corn was an activity important to the economy of the village.

Several of the rooms contained large stacks of corn laid up in recog-

FIGURE 2.21. Corn in Room 24.

nizable rows, usually in the corners of the structure. Room 24, which had been badly burned, produced many ears of corn stacked in what had originally been neat rows (Fig. 2.21). We also found shelled corn in great quantities. Much of the 20 bushels recovered surrounded the large fragmented El Paso ollas. (See Chapter 3 for a more detailed description.)

Although we did not find the original corn fields of the site, we believe that they occupied the flat areas to the north, east, and south of the site. Because the village was located near Deer Creek, the flood farming (or portage of water) of the small plots of corn is a reasonable conclusion.

SUMMARY

The Joyce Well site, located in Hidalgo County, New Mexico, was investigated in 1963. From the information recovered as a result of the excavation and analysis we concluded the following summary points.

First, the site was typical of the Animas phase settlements in the area (Kidder et al. 1949; McCluney 1965a). The village layout consisted of contiguous rooms surrounding a central plaza.

Second, the initial occupation of the site, based upon the correlation of pottery and past information from similar sites was made ca. A.D. 1250–1275. The abandonment, probably periodical rather than en masse, began about A.D. 1300–1350 and continued until final abandonment was completed about A.D. 1380–1400.

Third, from analysis of the total artifact assemblage of the site we conclude that the inhabitants were influenced by Casas Grandes peoples. We did not, however, observe any of the concentrated systems of manufacturing or trade displayed at the Casas Grandes site.

Fourth, the people of the El Paso phase were in contact with and exchanged with the people of Joyce Well.

It remains for the future investigators of the sites of the Animas and Hachita valleys to seek the answers to the many questions that have arisen. The archaeology of the area is complex, and much time is necessary to bring together the scattered details to complete the archaeological picture.

ROOM DESCRIPTIONS

In this section we present more detailed information about the 45 rooms excavated at the Joyce Well site. Room numbers are in consecutive order in sequence of excavation. The descriptions, which come from room excavation forms, are followed by the individual room drawings.

The room dimensions were taken along the inside surfaces of the room walls. All rooms were excavated to floor level, with occasional test holes through the floor. Wall lengths are those at the time of excavation. Coordinates of feature locations are reported from a common datum point, the inside southwest corner of each room.

Drawings are from the Eugene McCluney Archive of the School of American Research Hidalgo County Survey in the Collections of the Archive of the Laboratory of Anthropology, item 9, 29HISAR63-16.

ROOM 1

Dates Worked:	June 11–14, 1963.
Form:	Almost square.
Dimensions:	North wall, 3.85 m; east wall, 3.06 m; south wall, 3.66 m; west wall, 2.94 m. Depth of excavation was 15–20 cm, from contemporary surface to top of walls.
Walls:	Adobe, built up in coursed, horizontal layers, was visible at times where plaster was missing. Surface of wall left pebbly, perhaps to bind plaster. Occasional vertical shrinkage cracks were visible. South wall was 55 cm high, weathered to 31 cm on west and north. Plaster was made of sifted adobe, about 4 cm thick, applied in multiple layers and smoothed on surface with brush or stone. Shrinkage cracks were common. The plaster spalled on drying after excavation.
Entrances:	Rectangular doorway in north wall was sealed with adobe and stones and plastered over from Room 1 side. Opening was 24 cm wide and 23 cm higher with sill 16 cm above floor. East edge 1.78 m from northeast corner of room. Open doorway in east wall, 24 cm wide and 25 cm high, with semicircular base 14 cm above floor. Doorway, 1.31 m from southeast corner of room, opens into Plaza 2. In south wall, sealed door-

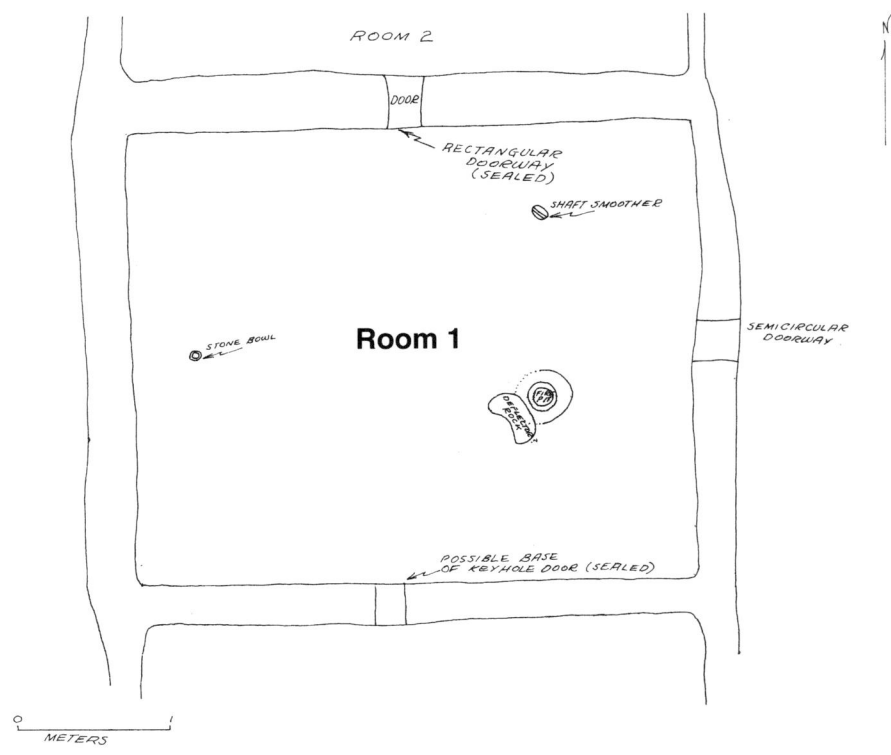

way indicated by rectangular block suggesting base of T-shaped door. Opening 20 cm wide, 25 cm high, and 20 cm above floor. Lower left corner 1.96 cm from southeast corner of room.

Floor:
The floor was slightly uneven, plastered with multiple layers made from sifted adobe. The floor was smoothly joined with wall plaster. Shrinkage cracks were visible over most of surface.

Roof:
Only evidence of the roof was fragments of charcoal and reed-marked adobe.

Features:
A circular firepit, 21 cm diameter, was located at 1.37 m north from the south wall and 2.77 m east from the west wall. The pit was lined with multiple layers of plaster and surrounded to slightly above floor level by plaster collar, which was somewhat depressed on east side, perhaps for ash removal. Large kidney-shaped rock in southwest quadrant of collar was possibly used as a deflector. Four-sided clay plug, 12 cm high and near room center, may represent base of roof support. The clay plug was plastered smooth on all sides and measures 28 cm north-south and 29 cm east-west. It was located at 1.53 north from the south wall and 1.86 east from the west wall.

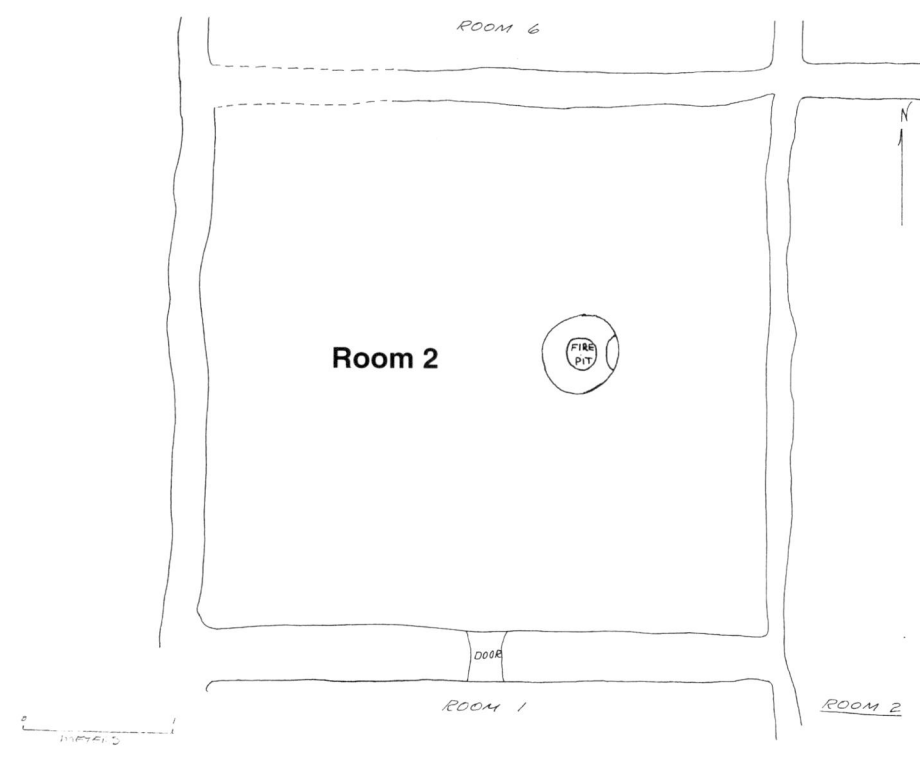

ROOM 2

Dates Worked:	June 13–15, 1963.
Form:	Almost square.
Dimensions:	North wall, 3.65 m; east wall, 3.69 m; south wall, 3.70 m; west wall, 3.49 m. Depth of excavation was 15–20 cm.
Walls:	Typical construction as in Room 1. South wall 38 cm high, weathered to 28 cm on west side. Wall thickness was 24 to 28 cm. Adobe courses visible in south wall indicate slumping toward west. Rodent holes and roots were present. Plaster preserved only 10–20 cm above floor. Walls curve to meet floor.
Entrances:	One sealed rectangular doorway in south wall, corresponding to north wall of Room 1. Any other entranceways possibly destroyed by erosion.
Floor:	Plaster in good to fair condition but had some evidence of rodent activity.
Roof:	There were scattered fragments of charcoal in fill at floor and some bonding of roof adobe to floor.
Features:	One 20-cm-diameter by 12-cm-deep firepit, at 2.56 m east of west wall and 1.85 m north of south wall. It had a collar of multiple layers of plaster, 50 cm in diameter, with an ash lip on east side. Possible 17-cm-diameter posthole at 1.19 m north of south wall and 2.05 m east of west wall indicated by soft, circular spot in floor. Depth was indeterminate.
Burial:	Present, but burial form is missing.

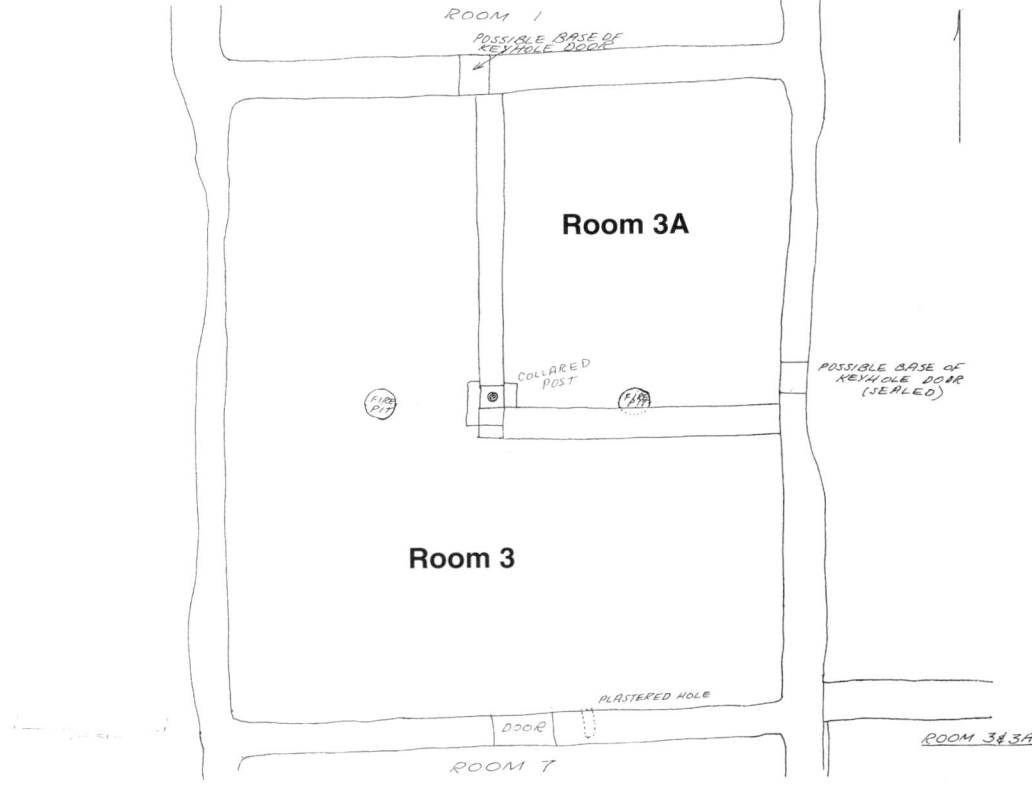

Room 3

Dates Worked:	June 12–23, 1963.
Form:	Almost square.
Dimensions:	North wall, 3.60 m; east wall, 4.18 m; south wall 3.62 m; west wall 4.09 m. Depth of excavation was 12 to 25 cm.
Walls:	The walls have the typical construction; 58 cm high eroded to 33 cm on west side; 22 cm thick. Adobe courses, 20 cm high, were visible in south wall. North and south walls were built into east wall. Vertical cracks were visible in south wall, 93 cm from southeast corner. Plaster is 4 cm thick and preserved to 35 cm above floor.
Entrances:	In north wall there was a sealed base of a possible T-shaped door corresponding to south wall of Room 1. Another doorway of same type is found in the east wall opening into Plaza 2. The doorway was sealed from Room 3. In south wall there is a rectangular doorway sealed from adjacent Room 7. Doorway measures 35 cm high and 41 cm wide, and the base is 14 cm above floor. Lower left corner is located 1.53 cm from southeast corner of room. Doorway is sealed with a large rock, oriented diagonally, and adobe.
Floor:	The floor had mostly intact plaster, but it was slightly irregular with evidence of rodent and root action.
Roof:	We found evidence of charred timber fragments, burned reeds and grass, and pieces of roof adobe on floor.

Features:	First firepit, located at 2.13 m north of the south wall and 0.89 m east of the west wall, was 21 cm in diameter and 13 cm in depth. Adobe collar was 38 cm in diameter with an ash lip on west side. Second firepit, at 2.04 m north of the south wall and 2.64 m east of the west wall, was 20 cm in diameter and 14 cm deep. Ash lip was visible on west side. South wall of Room 3A built over firepit, leaving 13 cm of pit exposed in Room 3A. There was a small, plaster-lined hole in south wall, 21 cm above floor. The hole is 5 cm in diameter and 18 cm deep, and located 1.30 m from southeast corner of room.
	A 7-cm-diameter timber post with a square adobe collar and pillar was found at 2.12 m north of the south wall and 1.65 m east of the west wall. Dimensions of collar: 39 cm north-south, 30 cm east-west, and 7 cm high. Column measurements: 19 cm north-south, 14 cm east-west, and 36 cm high. Adobe made from sifted soil as in plaster.
Burial:	Infant burial was found subfloor at center of west wall, almost under wall. Located 0.52 m from south wall and 1.13 m from north wall. Burial position is face up with the head to the southeast. White clay lines the bottom of the pit and a single Tucson Polychrome-like sherd was found.

ROOM 3A

Dates Worked:	June 20–23, 1963.
Form:	Almost square.
Dimensions:	Within Room 3. Depth of excavation was 25 cm.
Walls:	Walls were typical, although thinner than Room 3. Height of walls were 58 cm, and thickness was 18 cm. North-south wall rests on collar and butts against column of collar and post. East-west wall contacts column with south face of wall projecting beyond edge of collar.
Entrances:	There was a possible base of a T-shaped door located in the east wall in the extreme southeast corner of the room.
Floor:	Floor was plastered, and was in good condition.
Roof:	See Room 3.
Features:	Partially covered firepit, see Room 3. A storage pit, 26 cm diameter and 36 cm deep, was found at 0.47 m north and 1.68 m west of the southeast corner of the room.
Artifacts:	Broken metate was found in room fill.

ROOM 4

Dates Worked:	July 26, 1963.
Form:	Rectanguloid.
Dimensions:	North wall, 0.96 m; east wall, 1.53 m; south wall, 0.98 m; west wall, 1.54 m. Depth of excavation was 20 cm.
Walls:	Walls had the typical construction and were from 22 to 35 cm thick (west wall). Maximum wall height is 52 cm and plaster was preserved to about 40 cm above the floor. The south and north walls both abut the

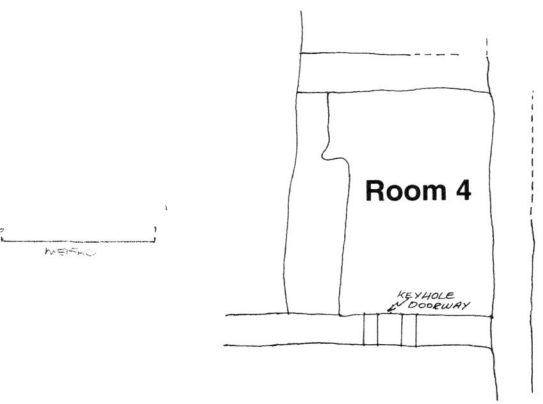

east wall and the west wall abuts the south wall.

Floor: The floor is plastered and is at the same level as the floor in Rom 14.

Entrances: The south wall has a sealed keyhole door about 52 cm from southeast
                  corner. It is 35 cm wide at the top and 17.5 cm at the base. The door
                  connects to Rom 14 and is sealed with plaster from both sides.

ROOM 5

Dates Worked: June 15–18, 1963.

Form: Almost square.

Dimensions: North wall, 3.19 m; south wall, 3.16 m; west wall, 3.11 m; east wall
                  3.00 m. Depth of excavation was 10–15 cm.

Walls: Walls were of the typical construction as described in Room 1. The
                  height of the walls was 32 cm, and the north and south walls were 26
                  cm thick and the west wall was 20 cm thick. The condition of the walls

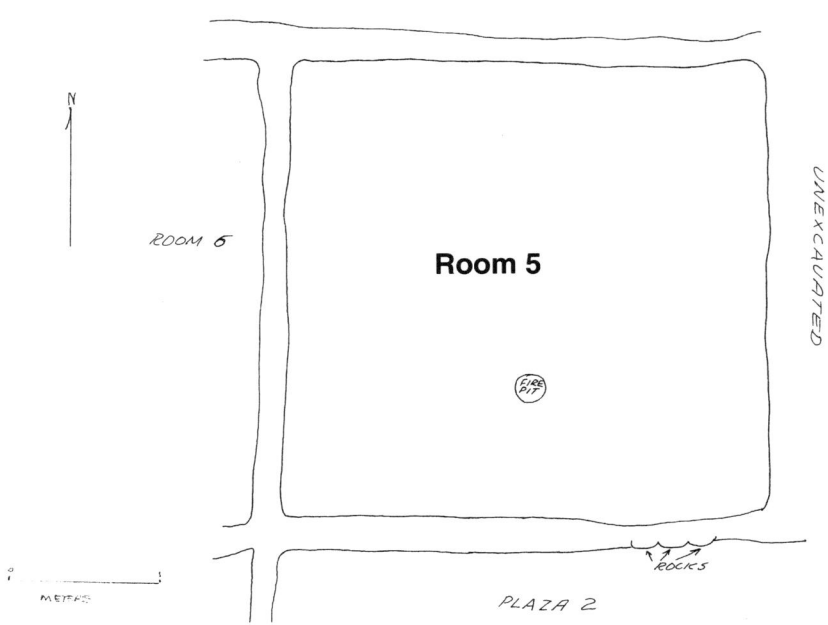

was generally poor as there was much evidence of erosion, root and rodent activity.

Entrances: None visible but they would have been destroyed by the disturbance.

Floor: The floor was uneven, poorly defined, and heavily disturbed. The presence of plaster could not be determined. There was some fallen roof adobe bonded with the floor.

Features: There is a circular fire pit with a diameter of 26 cm located at 80 cm north of the south wall and 1.50 m west of the east wall. The depth of the pit is 10 cm and there was not a distinct collar present, although there was a possible ash lip on the west side. There was a pile of stones in the northwest corner with no apparent purpose.

ROOM 6

Dates Worked: June 15–17, 1963.

Form: Rectangular.

Dimensions: North wall, 3.55 m; south wall, 3.71 m; west wall, 2.97 m, east wall 3.08 m. Depth of excavation was 8–20 cm.

Walls: Walls had the typical construction and reached a height of 42 cm on the south and 33 cm on the north. There was uniform erosion on all walls but no plaster was found.

Entrances: None were visible but any entrances would have been destroyed by disturbance.

Floor: Consisted of hard-packed earth and was even and pebbly.

Roof: No evidence of the roof except there was some ash, 1.42 m from northeast corner, along the north wall.

Features: A test pit was excavated into the floor and sterile soil was encountered.

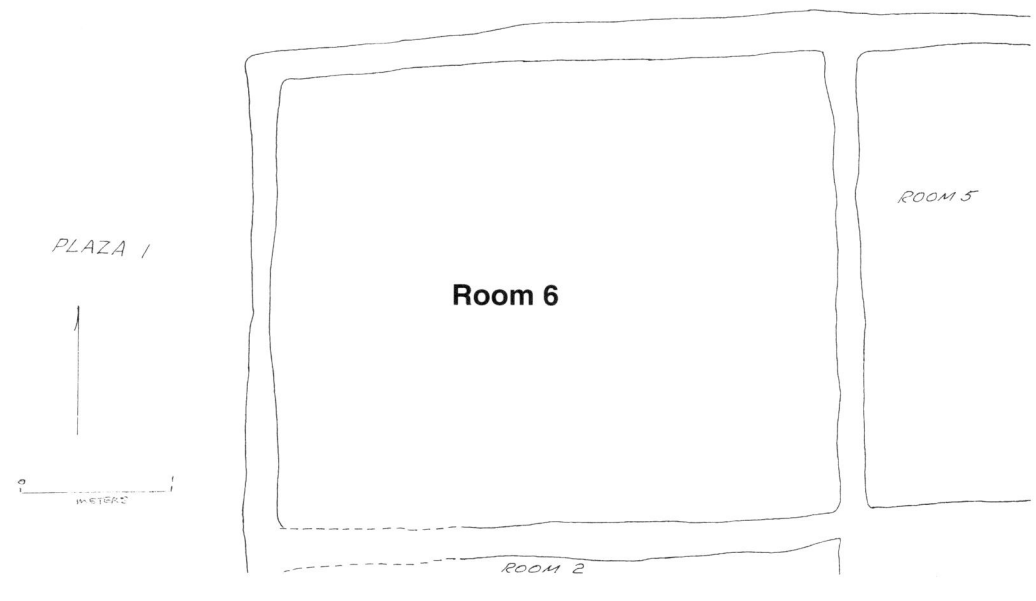

PLAZA 1

ROOM 5

Room 6

ROOM 2

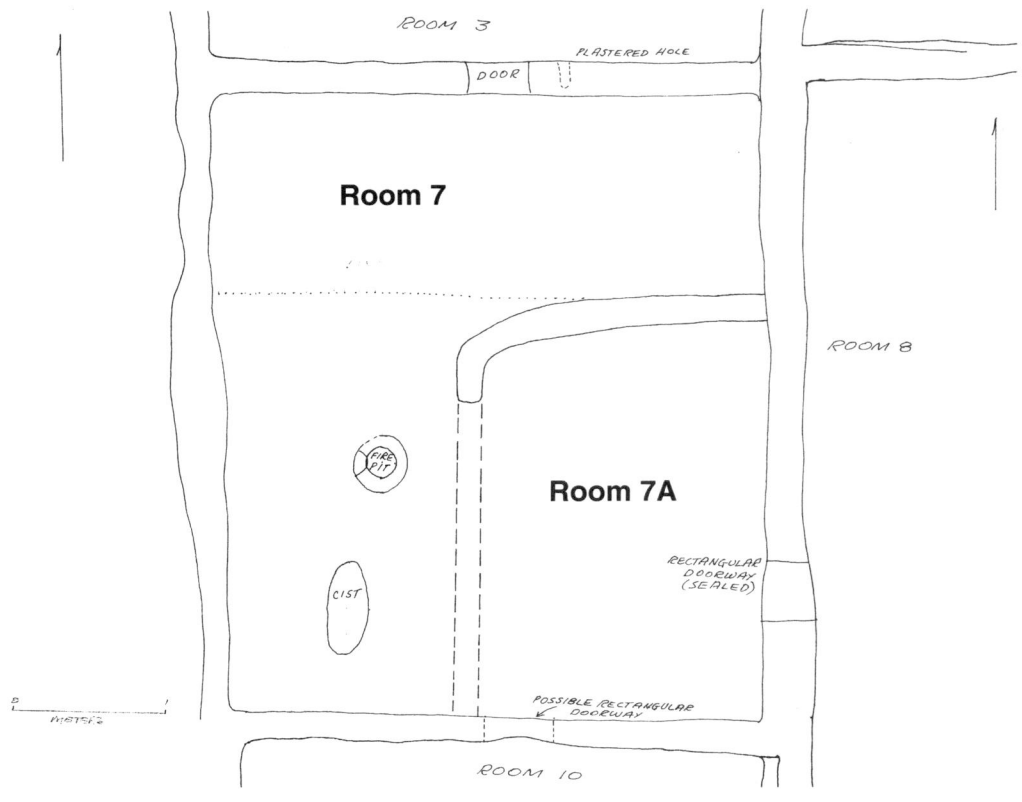

Room 7 and 7A

Dates Worked:	June 19–20, 1963.
Form:	Rectanguloid.
Dimensions:	North wall, 3.59 m; east wall, 4.27 m; south wall, 3.43 m; west wall, 4.19 m. Depth of excavation: 20 cm.
Walls:	The walls are of typical construction; 51 cm high, eroded to 20 cm on west side; 24 cm thick. Vertical joints were visible in north wall, which abuts east wall. Plaster preserved near floor over most of room, which rises to 34 cm on east side.
	Room 7A enclosed by poorly preserved partition in southeast quarter of room. Partition wall abuts east room wall at 1.71 m north and may have extended to west wall. Artifacts found south of extended wall line assigned to Room 7A. The standing partition was 2.67 m long east-west, 1.88 m south, and thickness was 16–18 cm.
Entrances:	A square doorway in north wall, sealed from Room 7, was common to Room 3 (south wall). In the east wall, there was an indistinct outline of a sealed doorway. There was also possible rectangular doorway in the south wall, 1.36 m from southeast corner of Room 7A. It is 46 cm wide.
Floor:	Typical plaster was in good condition, but was slightly irregular.
Features:	A circular firepit, 20 cm diameter and 14 cm deep, was found at 2.20 m north of south wall and 0.898 m east of west wall, with 8-cm-wide

adobe collar showing ash lip on west side. A niche, 8 cm diameter and 5 cm deep, with plastered interior, was found at floor level 2.30 m north of southeast corner of room. A possible cist, 28 cm east-west by 64 cm north-south, was located at 0.71 m north of south wall and 1.17 m east of west wall.

Burial: Room 7A had a subfloor extended burial lying face up along the east wall (11 cm from east wall). Head faces south but other information is missing on burial form. Twenty to twenty-five beads found around the neck and one bowl and one jar placed near the head. A bone needle, seeds, and a stone with three drill holes found near pelvis area.

A second burial was found in Room 7A but it is represented by just a few bone fragments. Burial seems to have been removed. Located sub-floor in southwest corner, 0.68 m north of south wall and 3.09 m west of east wall. Several shell beads were also found in pit.

ROOM 8

Dates Worked: June 21–25, 1963.
Form: Rectanguloid.
Dimensions: North wall, 3.42 m; east wall, 5.13 m; south wall, 3.54 m; west wall, 5.19 m. Depth of excavation, 15–20 cm.
Walls: Walls have the typical Joyce Well coursed adobe construction. Height of the north wall is 62 cm, which was eroded to 37 cm in north. Thickness was 24 cm. Vertical cracks were visible in the south wall. Rocks used in exterior base of wall in northeast corner. West wall abuts northeast corner of Room 10, 77 cm north of southwest corner. Plaster preserved to 40 cm above floor.
Entrances: Rectangular doorway in west wall, 39 cm wide and 12 cm above floor, sealed from Room 8 side. Lower left corner of the doorway is 1.48 m from southwest corner of room. In north wall there was a sealed rectangular doorway that once opened to Plaza 2. Opening is 32 cm wide and 8 cm above floor. There was also a rectangular doorway in the south wall, which connects to the north wall of Room 9.
Floor: Floor was typical plaster, although largely missing, which is on same grade as Level 1 of Plaza 2. A test hole, 1-by-1 m, in center of room showed Level 2 of Plaza 2 passes 14 cm below room floor.
Roof: Roof adobe was bonded unevenly to floor.
Features: A circular firepit, 21 cm in diameter and 14 cm deep, was located at 2.64 m north of south wall and 2.38 east of west wall. The firepit had an adobe collar, 42 cm north-south and 48 cm east-west, and an ash lip on the east side. A possible post, 17 cm in diameter, was also discovered at 1.67 north of south wall and 1.75 east of west wall and was indicated by soft spot in floor.

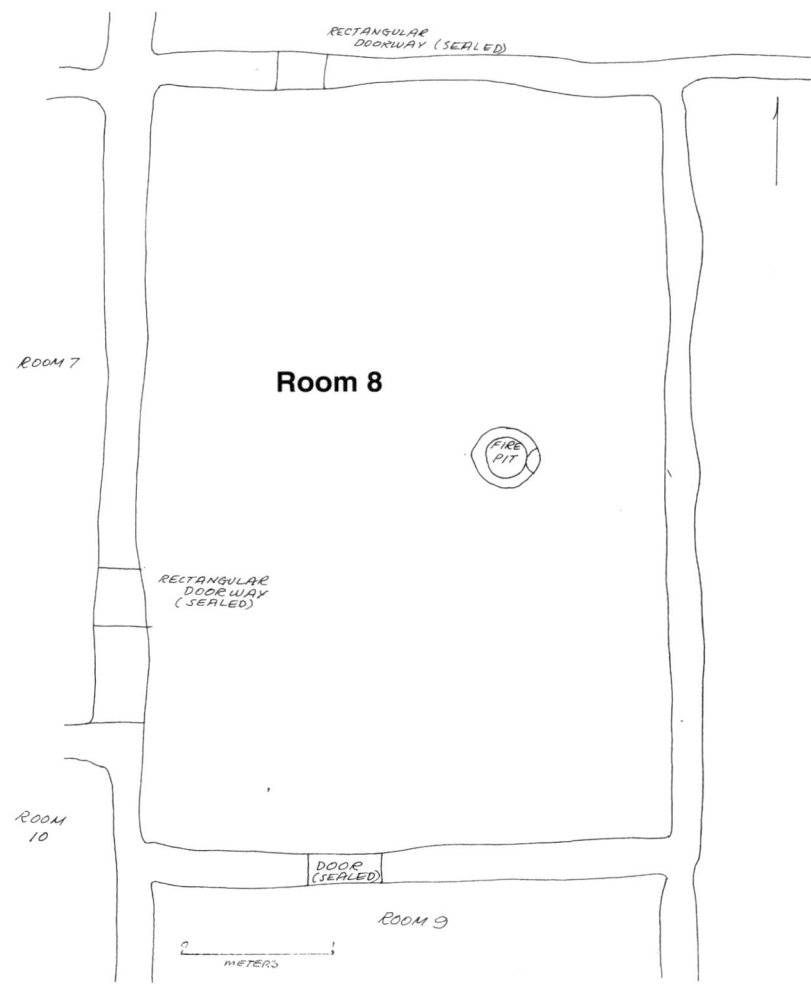

ROOM 9

Date Worked:	June 25–27, 1963.
Form:	Rectanguloid with nook in southwest corner.
Dimensions:	North wall, 3.33 m; east wall, 3.69 m; south wall, 2.45 m with 28 cm offset to south in nook, extending 78 cm to west; west wall, 3.77 m.
Walls:	Walls have typical construction and stand 66 cm high and 22 cm thick. East and west walls abut north and south walls. Remains of plaster to top of wall burned to orange and black colors. Wall behind plaster also burned orange to 2 cm depth.
Entrances:	In the north wall. There was a sealed rectangular doorway, 48 cm wide and 9 cm above the floor. The lower left corner of the doorway is 1.64 m from northwest corner of room. There was also a sealed T-shaped door in the east wall. (Lower right corner is 1.74 m from southeast corner of room.) Width of upper doorway is 60 cm and lower section is 18 cm wide by 15 cm high (base 9 cm above floor). In the south wall there

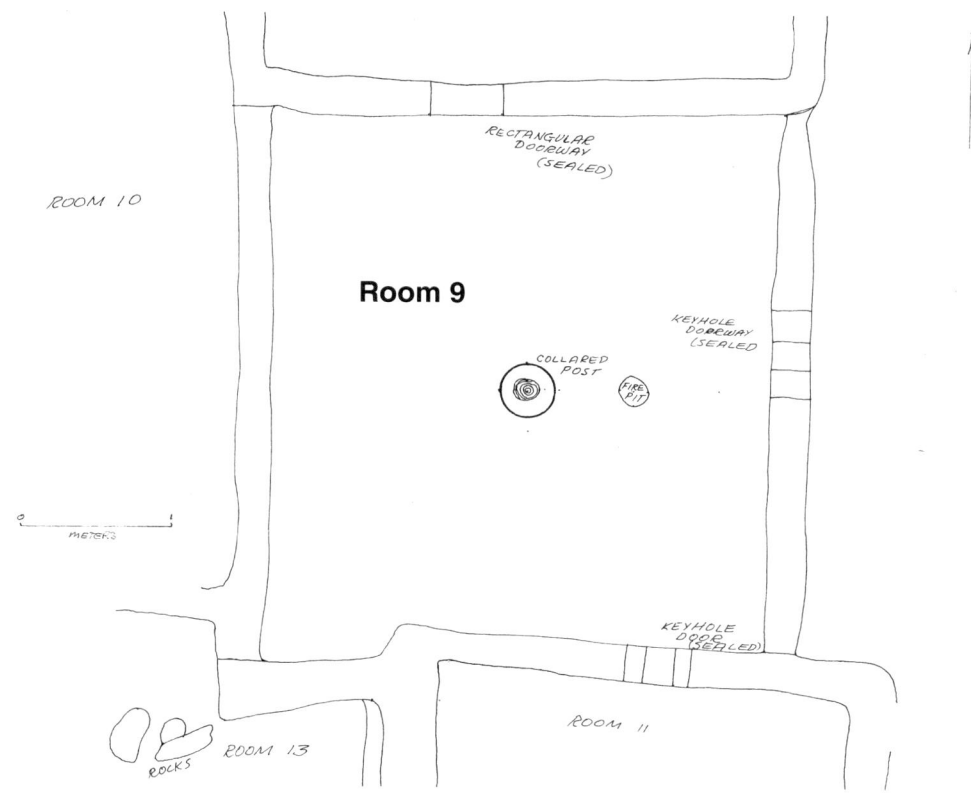

was a sealed T-shaped door that connects to Room 11 through its north wall.

Floor: Floor had typical construction, but plaster was cracked and burned to orange and black colors. Floor was buckled and slumped.

Features: A circular firepit, 20 cm in diameter and 13 cm deep, was located at 1.77 m north of south wall and 2.37 east of west wall. It had a 16 cm collar but no ash lip. A collared post, 16 cm diameter, was discovered in center of room at 1.69 north of south wall, 1.66 east of west wall. The adobe collar is 58 cm in diameter and 9 cm high. The stump of the carbonized post was visible.

Remarks: Masses of burned corn on the cob uncovered in northern half of room. Burned corn was first encountered in room fill 10 cm below top of wall.

Burial: Subfloor burial found along the west wall, 2.30 m west of east wall and 1.56 m north of south wall. Burial is face up with head to east. A Tucson Polychrome bowl, turquoise beads, and few sherds were also found.

ROOM 10

Dates Worked: June 24–26, 1963.
Form: Almost square.
Dimensions: North wall, 3.37 m; east wall, 4.04 m; south wall, 3.26 m; west wall, 4.01 m. Excavated to 20 cm depth.

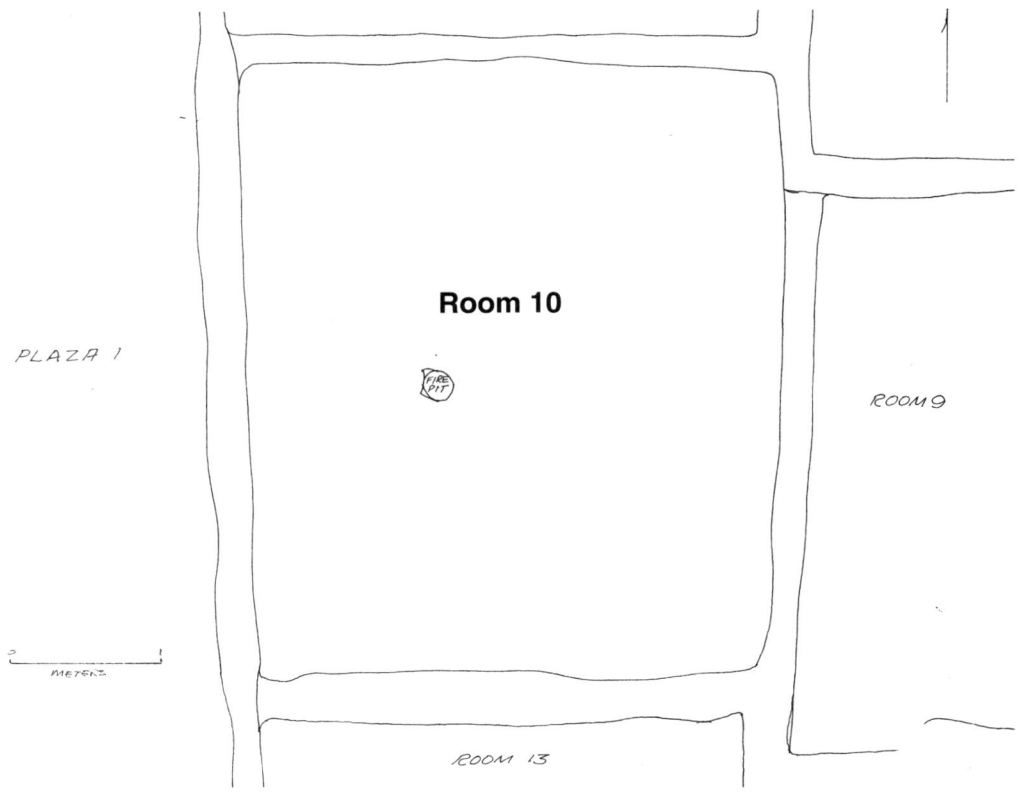

Walls:	The walls have the typical construction and were 24 cm thick. The maximum height of the wall was 56 cm, eroded to 26 cm at west side. North and south walls abut west wall. North end of east wall abuts northwest corner of Room 9. Plaster preserved to 20–30 cm above floor, which was burned to black and orange color and brittle.
Entrances:	None.
Floor:	The floor had the typical multilayered plaster but was uneven. Most of floor preserved was but is extensively damaged by rodents.
Roof:	Some charcoal was encountered in the fill at the floor, but no pieces were large enough to indicate orientation or type of roofing materials.
Features:	Circular firepit, 20 cm diameter and 12 cm deep, located at 1.92 north of south wall and 1.02 m east of west wall. Ash lip on the west side.
Artifacts:	Playas Red bowl and floor polisher on floor.
Burial:	A burial pit was found 1.60 m from northeast corner and 0.35 m west of east wall. Bones have been removed from pit. Two shell beads and some sherds were found.

ROOM 11

Dates Worked:	June 27–28, 1963.
Form:	Rectanguloid.

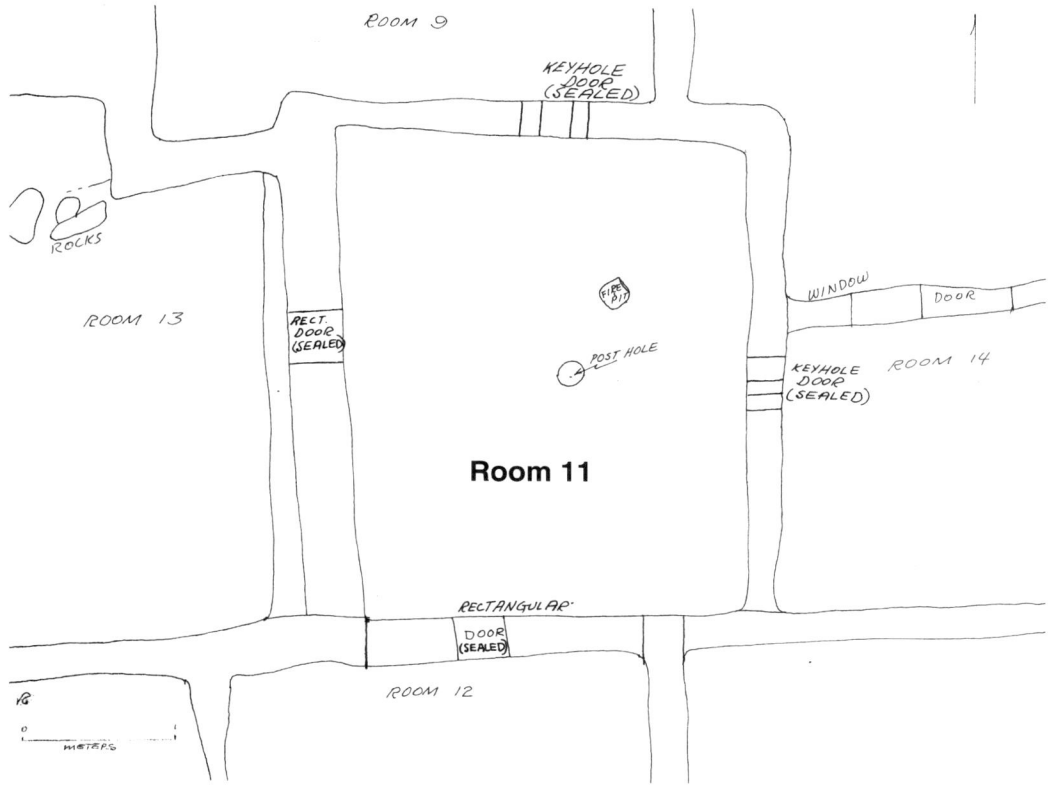

Dimensions:	North wall, 2.72 m; east wall, 3.16 m; south wall, 2.51 m; west wall, 3.46 m. Excavated to 15 cm depth.
Walls:	Walls had the typical construction and were 23 cm thick generally, but 33 cm thick at base of the west wall. East wall abuts south wall, common to Room 15 and west wall abuts south wall where common to Room 12. Plaster preserved to 42 cm above floor.
Entrances:	This room has a doorway in each wall. A sealed T-shaped door in the north wall was plastered over from both sides. The upper section of the doorway is 42 cm wide, and the lower portion is 23 cm wide, 24 cm high, and its base is flush with the floor. Lower left corner of doorway 1.26 m from northwest corner of room. The east wall has a sealed doorway common to the west wall of Room 14. A sealed rectangular doorway occurs in the south wall and is 32 cm wide and 27 cm above floor. The lower right corner is 56 cm from southwest corner of room. A doorway also was found in the west wall that connected (when unsealed) to Room 13.
Floor:	The floor had typical plaster and was slightly uneven.
Features:	An almost square firepit with rounded corners was found at 2.19 m north of south wall and 1.86 east of west wall. It measures 20 cm northeast-southwest, 18 cm northwest-southeast, and 8 cm deep. The firepit's ash lip is on the northeast corner. A posthole, 18 cm in diameter, was

found at 1.62 m north of south wall and 1.58 east of west wall, but its depth is indeterminate.

Burial: An infant burial was found subfloor in northwest corner of the room. Burial is on back and head is to the east. A projectile point found near skull, and corn cobs and charcoal found in pit.

ROOM 12

Dates Worked: June 31–July 2, 1963.
Form: Rectanguloid.
Dimensions: North wall, 2.72 m; south wall, 2.65 m; west wall, 3.60 m; east wall 3.64 m. Depth of excavation was 20–25 cm.
Walls: Typical construction with a thickness of 24 cm. North wall had a height of 56 cm and the west wall was eroded to 45 cm. Some root action was noted. Plaster was preserved on the lower 30 cm of each wall. North wall abuts into the east wall.
Entrances: There are three doorways. The south wall has a sealed doorway that connects with Room 20, the east wall has a sealed doorway that connects with Room 15, and the north wall has a sealed door that connects with Room 11.
Floor: Plaster is visible but irregular and is in good to poor condition.

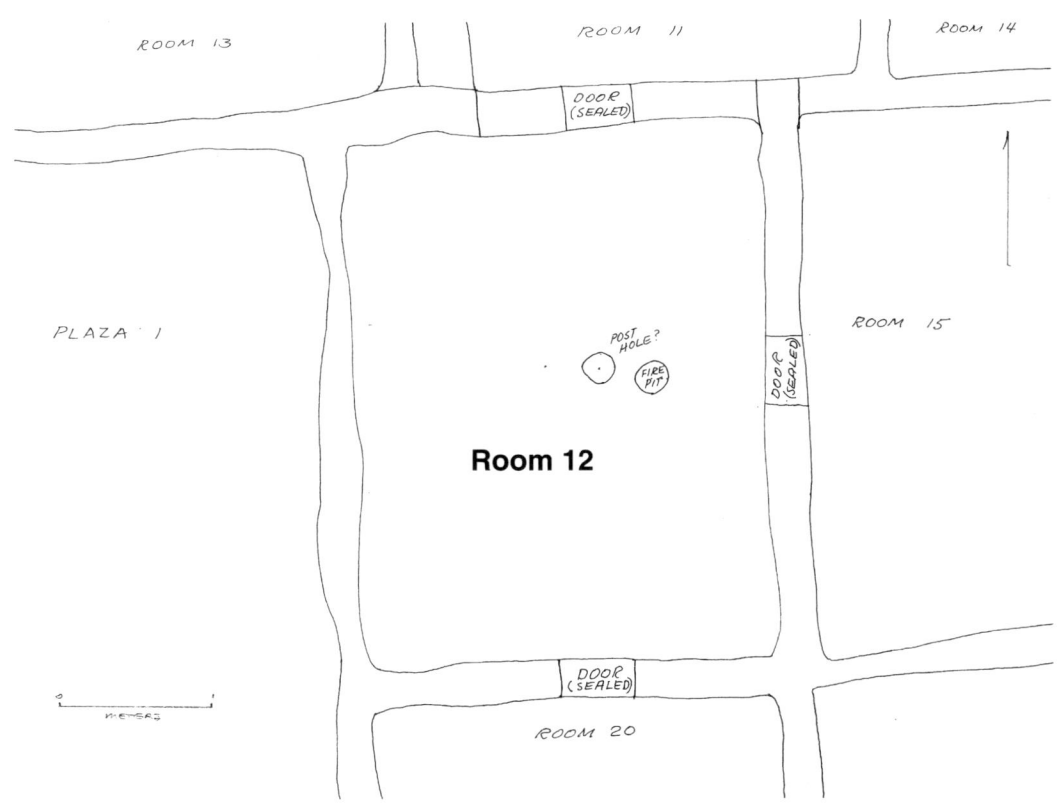

Features: A circular, poorly preserved fire pit was found with a diameter of 22 cm and an unknown depth. It was located at 1.88 m north of south wall and 75 cm west of east wall. A possible posthole was encountered at 1.97 m north of south wall and 1.07 m west of east wall and had a diameter of 20 cm.

Burial: A subfloor burial was in the northwest corner of the room. Burial lies on its left side (facing south) and head is to east. Mouth is filled with juniper berries.

ROOM 13

Dates Worked: June 29–30, 1963.

Form: Rectanguloid, with intrusion of Room 9 walls into northeast corner.

Dimensions: North wall, 3.18 m to offset, then 44 cm south, then another 1.02 m to east wall; east wall, 3.3 m; south wall, 4.8 m; west wall, 3.69 m. Offset part of north wall extends 19 cm into room, terminating in a cluster of rocks. Excavated to depths of 20–25 cm.

Walls: Walls have the typical construction and had a maximum height of 64 cm, which eroded to 40 cm on west wall. Wall thickness was 24 cm. Some rodent action was visible. North wall abuts west wall and east wall abuts south. On north, east, and south walls, an older wall is visible at lower level. On east side, older wall extends 14 cm into room forming a shelf 32 cm above floor. There was some intact plaster on lower portion of east wall only.

Entrances: There was a rectangular doorway in east wall, sealed from this room, 42 cm wide and 0.90 m from northwest corner. Door sill was reused as a shelf.

Floor: Plaster was in good condition but slightly irregular with some slumping and bare spots. On a subfloor, 8 cm below the plaster surface, ash and rocks are found at plaza level.

Features: This room had two firepits. An oblong firepit, 20 cm east-west, 23 cm north-south, and 12 cm deep, was located at 1.8 m north of south wall and 2.5 m east of west wall. The ash lip for the firepit was on south edge. A second firepit was circular, 19 cm diameter and 11 cm deep, and was located at 2.74 m north of south wall and 2.63 east of west wall. The ash lip was on the north edge.

ROOM 14

Dates Worked: July 2–3, 1963.
Form: Rectanguloid.
Dimensions: North wall, 6.63 m; east wall, 1.93 m; south wall, 6.80 m; west wall, 1.97 m. Excavated to a depth of 16 cm.

Walls:

Wall construction identical to other rooms. Maximum height was 62 cm, which eroded to 52 cm on east side. Thickness 22 cm. On the west wall layering was visible with the course thickness about 31 cm. North wall abuts east and west wall. North wall of Room 16 interrupts south wall.

Entrances:

Five doorways were found in this room. On the west end of the north wall, a rectangular doorway was found that had been closed with soft fill. The left corner was 1.44 m from northwest corner of room. The door opening was 53 cm wide and 44 cm high; sill 18 cm above floor. On the east end of the north wall, a sealed T-shaped doorway opened into Room 4. The entryway was plastered over from both sides. The lower right corner was located 52 cm from northeast corner of room. The upper opening was 35 cm wide, and the lower 17 cm high by 17.5 cm wide and 14 cm above floor. On the east side of the south wall a rectangular doorway was found that had been sealed with large rocks from the Room 14 side. The left side of the doorway abuts east wall of Room 16, 1.75 m from southeast corner of Room 14. This opening is 80 cm wide and flush with the floor. On the west side of the south wall a T-shaped door closed with soft fill. The lower left corner of this doorway is located 1.24 m from southwest corner of room. Upper portion of the opening was 48 cm wide, and the lower section was 21 cm wide by 26 cm high; sill was 6 cm above floor.

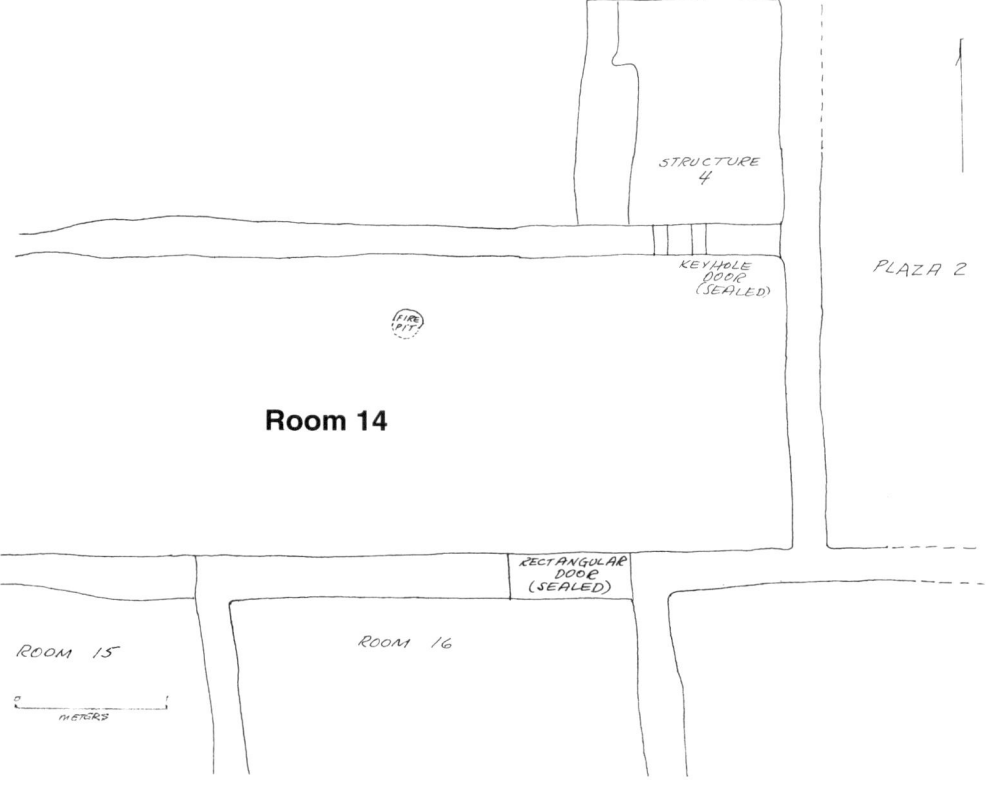

Floor:	The plaster surface was warped and irregular over most of room.
Roof:	Roof plaster was bonded to floor.
Features:	A partially destroyed firepit was discovered at 1.56 m north of south wall and 4.11 m east of west wall. Its diameter was about 19 cm, but the depth was uncertain. An oblong cist, 24 cm east-west and 21 cm north-south, was located at 0.63 m north of south wall and 0.69 m east of west wall. It had a depth of 21 cm, and it sloped to the south. A hole in the floor was identified at 0.19 m north of south wall and 1.62 m east of west wall. It was a possible animal burrow leading under south wall to Room 15. Dimensions of the hole were 72 cm east-west by 37 cm north-south, and 41 cm deep. Fill included sherds and charcoal. A window was also identified in the northwest corner of room, abutting the west wall. The window is 44 cm wide by 14 cm high.
Artifacts:	Two El Paso Polychrome sherds were found embedded in the north wall.
Burials:	A subfloor infant burial was found along west wall. The body lies on left side and faces west. The burial was partly under the west wall of the room. One paint tube and two turquoise pendants were found in association with the burial.
	This room also had a semiflexed, subfloor burial along the south wall. Body lies on back and head was to the west. A bowl was placed near the skull and a grass bundle possibly containing beans was just below the head and another grass container was located near the pelvis.

ROOM 15

Dates Worked:	July 5–13, 1963.
Form:	Trapezoidal.
Dimensions:	North wall, 3.63 m; east wall, 3.26 m; south wall, 3.83 m; west wall, 3.70 m. Excavated to average depth of 20 cm.
Walls:	Walls have usual construction and have a median height of 75 cm and thickness of 22 cm. North wall abuts east wall. Plaster was intact to 58 cm on south wall. An adobe course was visible at top of north wall.
Entrances:	The north wall had a T-shaped door, with an upper opening of 48 cm wide and a lower opening of 21 cm wide by 26 cm high; base was filled in to 18 cm above the floor. The lower left corner of T-shaped opening was 1.69 m from northwest corner of room. The east wall had a possible sealed door. The doorway in the south wall was common to Room 18's north wall. A sealed rectangular door in west wall was uncovered (1.49 m from northwest corner of room to lower right corner of doorway.) The dimensions of the doorway were 48 cm high by 49 cm wide with a sill 18 cm above floor. The opening was filled with rocks and adobe.
Floor:	Plaster in excellent condition. Sterile soil encountered below floor.
Roof:	Fragments of burned rafter, oriented north-south, found in room fill along with burned reeds.
Features:	A circular firepit, 20 cm diameter and 10 cm deep, was located at

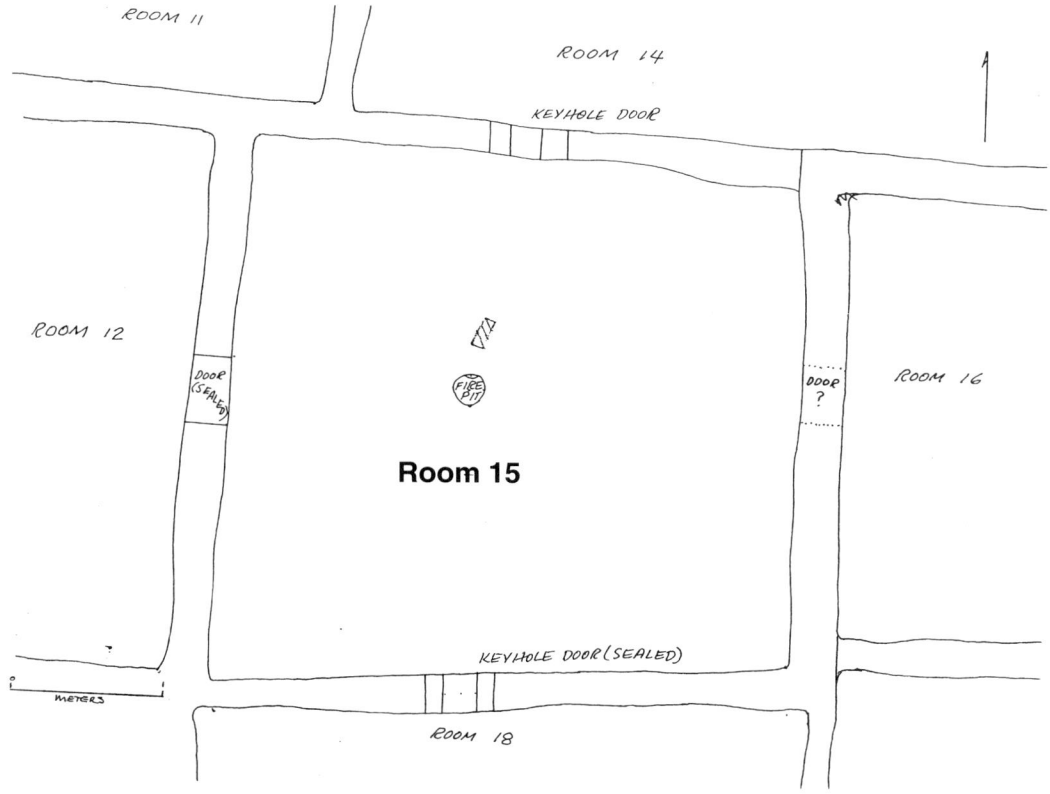

1.87 m north and 1.41 m east. The ash lip was found on the north edge of the firepit.

Artifacts: A crushed olla was found on the floor at 3.30 m north of south wall and 2.10 m east of west wall; floor polisher at 2.08 m north of south wall, 1.05 m east of west wall; pestle at 1.86 m north of south wall, 0.83 m east of west wall.

Burials: A flexed, subfloor burial was found along the west wall lying on its left side. The head is to the north and faces east. A red punctate bowl was inverted over a smeared corrugated jar, which contained a gourd dipper and corn meal. A cluster of seeds was found near the head and under the pelvis. Corn and corn cob also found.

   A second burial was found in the southeast corner of the room, 0.24 m north of south wall and 1.0 m west of east wall. The burial is lying on its left side, the head faces west. No sketch was possible because bones were fragmentary. Four strings of stone beads, grass matting or woven bundle under skull, and an effigy vessel were found. Cotton cloth, 16 strands per cm, found in several layers. Possibly used to cover the head. Frog pendant also found. Corn was found on north side and seeds under the body.

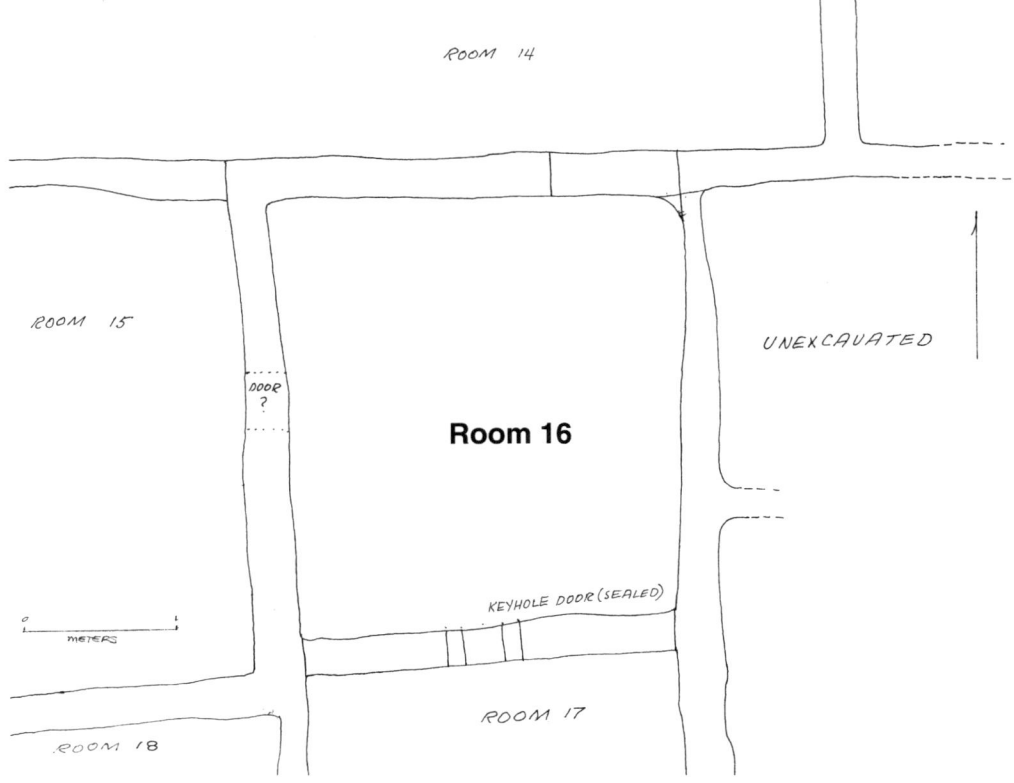

ROOM 16

Dates Worked:	July 9–10, 1963.
Form:	Almost square.
Dimensions:	North wall, 2.68 m; south wall, 2.50 m; west wall, 3.01 m; east wall 2.78 m.
Walls:	Typical construction to the walls that were 28 cm thick and 59 cm high. Plaster was preserved to 51 cm above floor and was burned to a black and orange color. The east wall abuts into the north wall, and the south wall abuts into the east and west walls.
Entrances:	The room has three sealed doorways. In the south wall is a keyhole door, 49 cm wide on top and 22 cm on bottom, which is located 1.07 m from southeast corner of the room to the lower southeast corner of the doorway. The doorway is sealed from Room 17 with rocks and adobe. The north wall has a rectangular doorway, flush with floor, that is located in the northeast corner. It is 80 cm wide and it was sealed from Room 14 with large rocks. The west wall has a sealed rectangular door but the size and shape are indeterminate because of disturbance.
Floor:	The floor was plastered but and was uneven, buckled and blackened, from the heat of the fire.
Features:	Approximately 11 cm below the floor a plaza-like surface was found that was separated from the room floor by ash and burned material.

Burials: Three burials were found. The first burial was located in the northwest corner, 14 cm from north wall and 6 cm from west wall. Burial was in the sitting position, facing northeast. Grave goods included a bead necklace, a Ramos Polychrome bowl, yucca fiber cord fragment adhering to a large sherd, and a black bowl that contained maize. A mat of yucca fiber below grave furniture. Large sherds also have bean or seed husks.

Burial 2 is found in the northeast corner, also in a sitting position facing southwest. A bowl and a necklace were also found in association.

A third subfloor burial is a child and is located in the southwest corner of the room. Body was on its back with the head to the south. A cord-impressed bowl, with a fabric mat underneath, is associated with the burial.

ROOM 17

Dates Worked: July 10, 1963.

Form: Rectanguloid.

Dimensions: North wall, 2.45 m; east wall, 6.76 m; south wall, 2.46 m; west wall, 6.78 m. Depth of excavation 19 cm.

Walls: Wall height was 63 cm, but was eroded to 50 cm on east wall. Thickness was 24 cm. North wall abuts east and west walls. Plaster was preserved to 48 cm above floor. Some root action was evident.

Entrances: In the south wall there was a sealed rectangular doorway, at 1.20 m from southeast corner of room. The doorway opens into Plaza 3 and has an opening of 53 cm wide; sill about 10 cm above floor. Doorway was sealed from room side. In west wall, south end, there was a sealed T-shaped doorway common to Room 19, east wall. On the west wall, north end, there is also a sealed T-shaped doorway, common to Room 18, east wall.

Floor: Floor was plastered and in excellent condition.

Roof: Only evident in small fragments of charcoal with no orientation.

Features: A nearly circular firepit was located at 2.89 m north of south wall and 1.04 m east of west wall. It measured 20 cm east-west, 22 cm north-south, and had a depth of 12 cm. Ash lip located on east side.

Burials: Burial 17/1 was in the southwest corner of the room. The subfloor burial was in a sitting position facing northeast.

Burial 17/2 was also located in the southwest corner and was in a sitting position facing west.

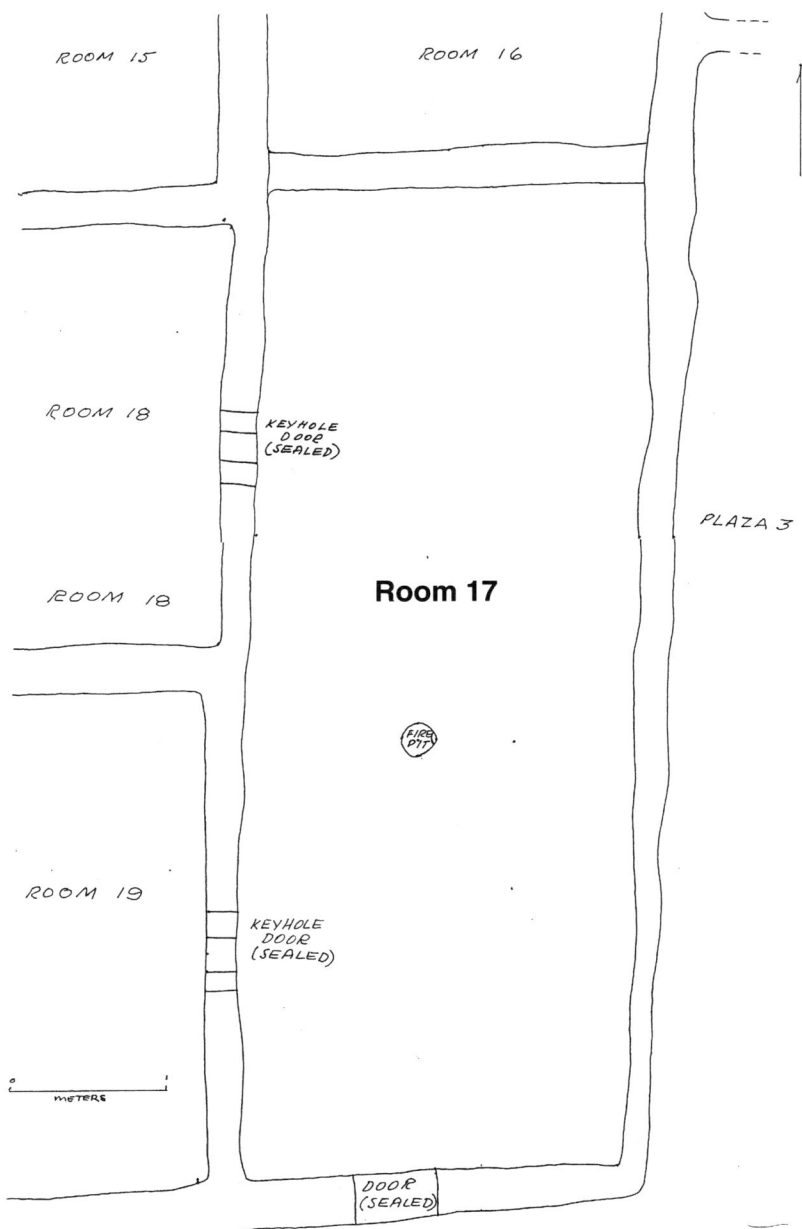

ROOM 18

Dates Worked:	July 11–15, 1963.
Form:	Rectanguloid.
Dimensions:	North wall, 3.84 m; east wall, 2.75 m; south wall, 3.84 m; west wall, 2.57 m. Excavated to depths of 15–20 cm.
Walls:	Walls have the typical coursed adobe construction. Wall height was 78 cm and thickness 24–32 cm. Plaster was 2 cm thick and preserved to top

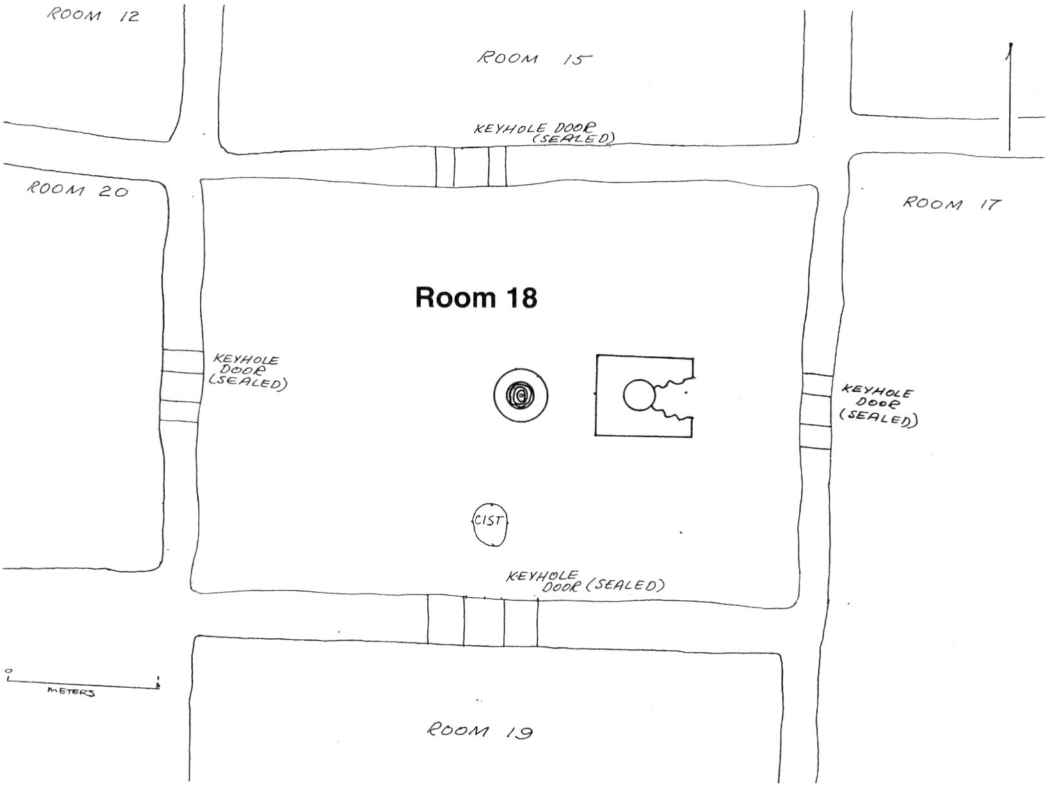

ROOM 12

ROOM 15

KEYHOLE DOOR
(SEALED)

ROOM 20

ROOM 17

Room 18

KEYHOLE
DOOR
(SEALED)

KEYHOLE
DOOR
(SEALED)

CIST

KEYHOLE
DOOR (SEALED)

METERS

ROOM 19

of walls. Plaster was burned to black and orange colors. Small amount of root and rodent damage was visible.

Entrances: On the north wall, 1.47 m from northwest corner of room, a base of T-shaped door was discovered. The base measured 43 by 23 cm and was 10 cm above floor and was sealed from both Rooms 15 and 18. In Room 18 the doorway was left partially open and used as a shelf. Another sealed T-shaped door was discovered in the east wall 1.27 m from northeast corner of room. Upper opening was 48 cm wide, lower 21 cm wide by 22 cm high, and the doorway base was 21 cm above floor. The doorway was filled from Room 17 with rocks, adobe, and a broken metate, and plastered over. The south wall had a sealed T-shaped door 1.71 m from southeast corner of room. Upper opening was 71 cm wide, lower

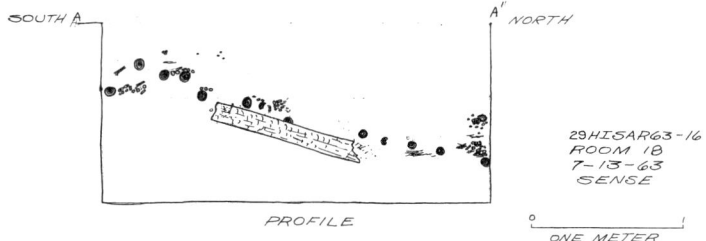

SOUTH A

A' NORTH

29HISAR63-16
ROOM 18
7-13-63
SENSE

PROFILE

ONE METER

	26 cm wide, and base was flush with floor. Opening was filled with rocks and adobe pieces from both sides. Doorway in west wall common to Room 20, east wall.
Floor:	Plaster was slightly irregular in places, but in excellent condition.
Roof:	Post, rafters, and cross rafters fallen to floor. Reeds, grass, and adobe plaster in position.
Features:	The room had a raised hearth, (1.31 m north of south wall, 2.85 m east of west wall) with 3-cm-thick rectangular adobe collar. The collar measured 59 cm east-west and 53 cm north-south. On the east side of collar was a tapered ash ramp, 29 cm at outer edge and 17 cm at inner edge; vertical sides of ramp were in scalloped form. Firepit was 25 cm in diameter, and 13 cm deep. The room also had a collared post, 1.31 north of south wall, 1.97 east of west wall, with a 17-cm-diameter post, 11 cm high, in situ. Adobe collar was 35 cm in diameter and 10 cm thick. Oblong cist was also located at 0.45 m north of south wall, 2.57 east of west wall. The cist measures 25 cm north-south by 23 cm east-west, and was 13 cm deep.
Burials:	Burial 18/1 was located against the west wall, 1.23 m north of south wall and 0.41 m west of east wall. Burial 18/2 was located along the south wall, 1.48 m west of east wall and 0.30 m north of south wall. No orientation was visible. Infant burial.

ROOM 19

Dates Worked:	July 14–16, 1963.
Form:	Almost square.
Dimensions:	North wall, 3.95 m; east wall, 3.33 m; south wall, 3.85 m; west wall, 3.49 m. Excavated to 29 cm depth.
Walls:	Wall height was 67 cm and thickness was 30 cm on the north, east, and south walls, and 24 cm on the west. South wall abuts east and west wall. Plaster preserved in places to top of wall. Some root and rodent damage.
Entrances:	In the east wall there was a sealed T-shaped doorway (lower left corner 1.48 m from northeast room corner) filled with rock and adobe from Room 17. Upper opening was 52 cm wide, lower 22 cm wide by 22 cm high, and base was 14 cm above floor. The south wall had a sealed rectangular doorway, 36 cm wide with sill 14 cm above floor (lower left corner 1.62 m from southeast room corner). Opening was plugged with adobe from Room 23 and the fill included El Paso Polychrome sherd. There was also a sealed doorway in west wall, common to Room 21, east wall, which was used as recessed shelf.
Floor:	Plaster construction in excellent condition. Test hole revealed another floor at about 35 cm depth. Intervening fill was very soft.
Features:	The room had a raised hearth of same construction as in Room 18, but northeast corner of collar fractured by roof fall. Found at 1.66 m north of south wall, 2.70 m east of west wall. Collar dimensions were 46 cm north-south, 50 cm east-west, and 3–4 cm high. Firepit was 22 cm in di-

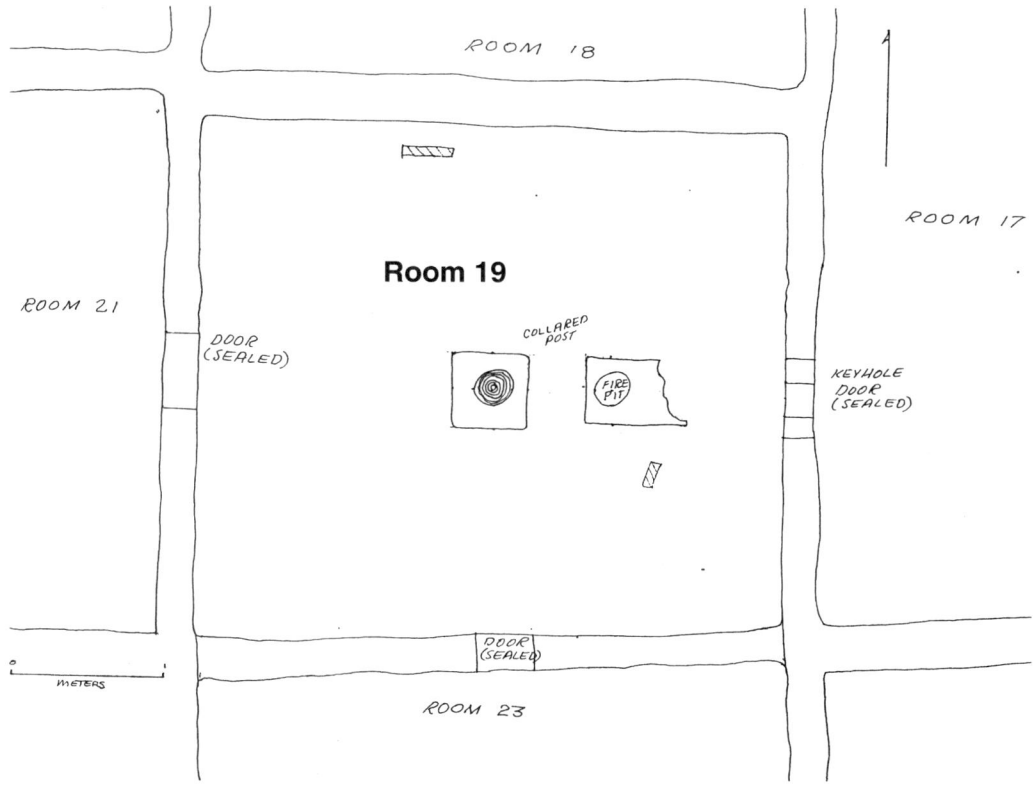

ameter and 10 cm deep. There was also a collared post, at room center (1.65 m north of south wall, 1.92 east of west wall) with an intact timber, 26 cm diameter and 1.10 m long. Collar measured 52 cm north-south, 50 cm east-west, and 67 cm high.

ROOM 20

Dates Worked:	July 15–17, 1963.
Form:	Near square, with intrusion in southeast corner.
Dimensions:	North wall, 2.69 m; east wall, 2.68 m; south wall, 2.0 m west to a 30-cm affect to north, continuing 74 cm to west wall; west wall, 2.55 m. Excavated to 15–20 cm depth.
Walls:	Walls had typical construction, and were 78 cm high but eroded to 50 cm on west. Wall thickness was 26 cm. North wall abuts east wall, and south wall abuts 55-cm-thick stub projecting 74 cm from west wall.
Entrances:	There was a sealed rectangular doorway in north wall, 49 cm wide and sill 23 cm above floor (lower left corner, 0.91 cm from northeast room corner). A T-shaped door in the east wall (lower left corner 1.14 m from northeast corner of room), was sealed with adobe from both sides. The doorway was 45 cm wide on the upper section and 20 cm wide on the lower; base 10 cm above floor. There was a possible sealed door in south wall, common with Room 21, north wall.

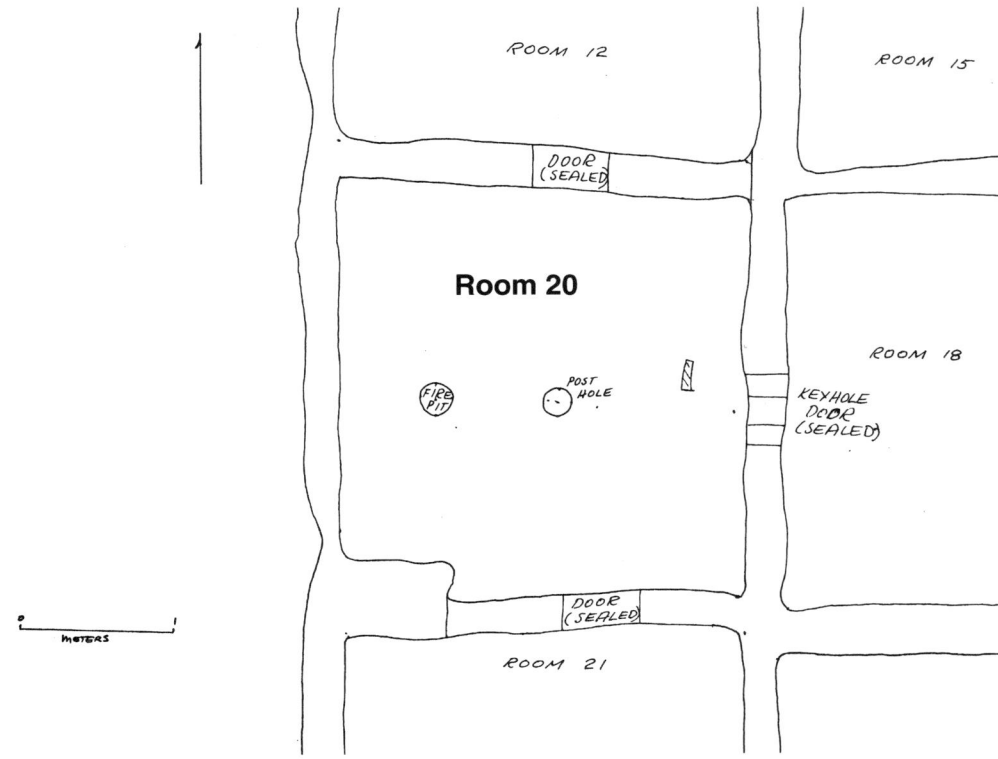

Floor:	Plaster construction in good condition.
Roof:	One fragment of a 5-cm-diameter burned rafter found in a north-south orientation.
Features:	A circular firepit in poor condition, 20 cm diameter, was found at 1.24 m north of south wall, 0.65 m east of west wall. A partially destroyed 18-cm-diameter posthole was located at 1.25 m north of south wall and 1.44 m east of west wall.

ROOM 21

Dates Worked:	July 16–17, 1963.
Form:	Rectanguloid.
Dimensions:	North wall, 2.63 m; east wall, 3.57 m; south wall, 2.65 m. Excavation depth 32 cm.
Walls:	Wall height was 61 cm but eroded to 54 cm at west wall. Wall thickness was 24 cm. South wall abuts east wall, north wall abuts west wall. There were vertical joints in north wall at each side of thick stub wall intruding into Room 20. Plaster preserved 10–20 cm above floor around room.
Entrances:	There was a sealed rectangular doorway in north wall, 49 cm wide with sill 23 cm above floor (lower right corner 65 cm from northeast corner of room). The opening was filled with rock and adobe and plastered from both rooms. A rectangular doorway in east wall was sealed from

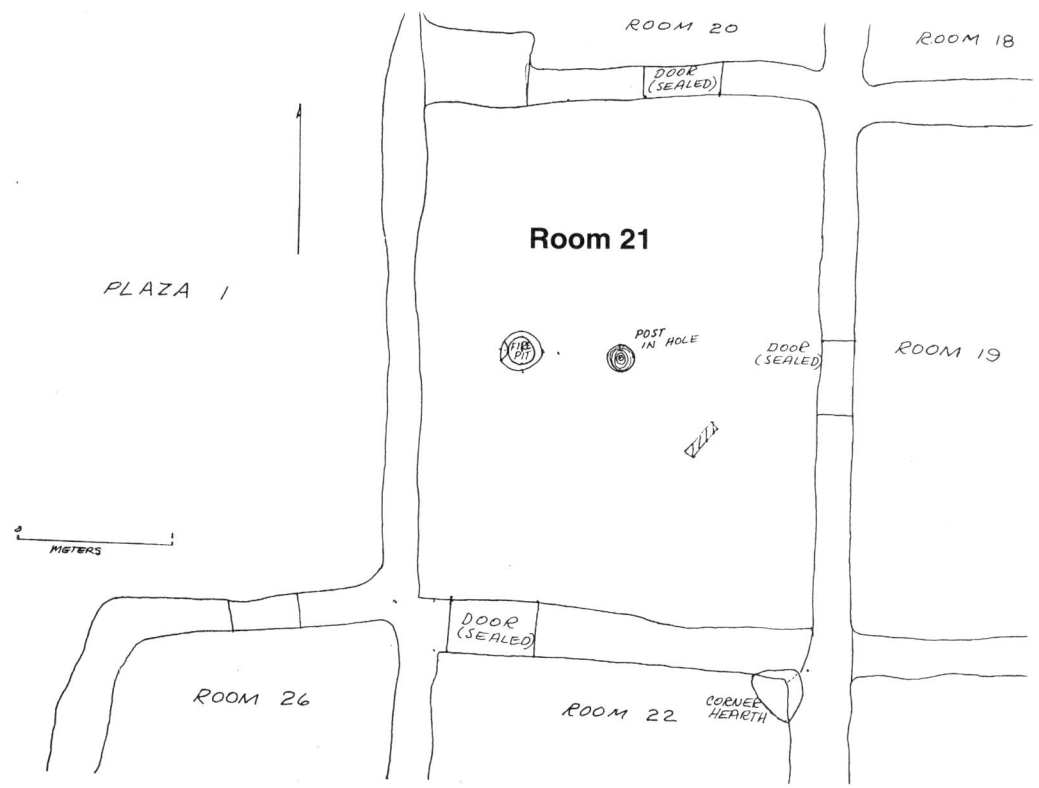

both sides with only 5 cm of adobe, but it was sealed flush on Room 21 side. In Room 19 the partially sealed opening was used as a recessed shelf. Opening 49 cm wide and sill was 25 cm above floor. The lower left corner was located at 1.58 cm from northeast room corner. In south wall, vertical joints perhaps indicate sealed, 57-cm-wide door, 20 cm from southwest corner of room.

Floor:	Plaster construction was in excellent condition, irregular in places.
Roof:	Two charred fragments of rafters oriented north-south. No evidence of ridge pole.
Features:	A circular firepit, 18 cm in diameter and 10 cm deep, was found at 1.77 m north of south wall, 0.68 m east of west wall. The firepit was enclosed in a 10-cm-wide adobe collar with the ash lip on west side. A burned, 17-cm-diameter post was found at 1.80 m north of south, 1.33 m east of west. Depth of the hole was indeterminate.
Remarks:	Abundant animal bones, charcoal, and sherds in room fill indicate room apparently filled with trash by occupants.
Burials:	Burial 21/1 was located near the southeast corner of the room, 1.40 m north of south and 0.49 m west of east. Burial is lying on left side, head is to the south and facing west. Gila Polychrome jar 10 cm west of skull and a Ramos Polychrome bowl included in grave. Clods of adobe impressed with fabric were in the fill. Bone tube filled with pigment also recovered. A heavy concentration of manganese was found on bones.

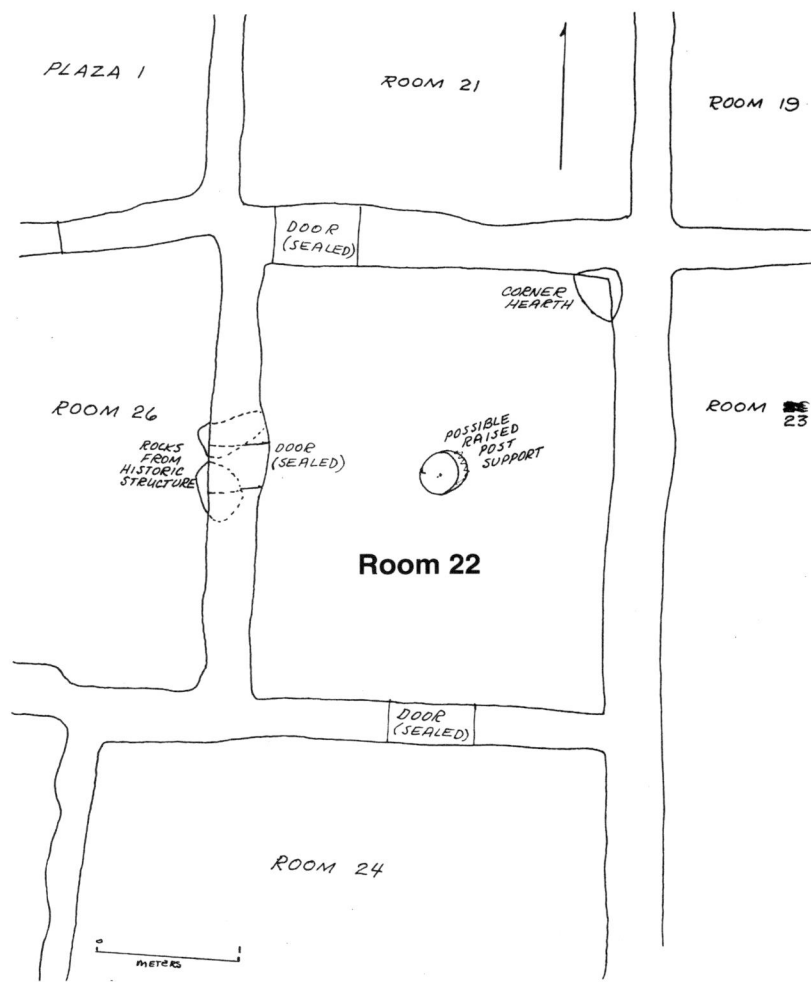

ROOM 22

Dates Worked:	July 19, 1963.
Form:	Rectanguloid.
Dimensions:	North wall, 2.32 m; east wall, 2.95 m; south wall, 2.36 m; west wall, 2.89 m. Excavated to depth of 31 cm.
Walls:	Walls had typical construction with height of 65 cm and thickness of 24 cm. North wall abuts east wall. Some damage to walls when historic house constructed over site. Rocks from house were encountered on west wall. Plaster preserved to 36 cm above floor.
Entrances:	A sealed opening in north wall connected with Room 21. Rectangular doorway in south wall sealed from Room 22. The opening was 55 cm wide and sill was 23 cm above floor (lower left corner 85 cm from southeast room corner). In west wall, there was a sealed doorway 28 cm wide and flush with floor, 1.24 m from northeast corner of room.
Floor:	Floor had typical plaster and it was in good condition.
Features:	A fire hearth was built into northeast corner of room with 2–3 cm of

adobe facing. Elliptical wings were observed against wall each 24 cm wide and 18 cm high. There was a possible raised support for post in center of room, at 1.53 m north of south wall, 1.17 m east of west wall. The support was made of adobe and was 22 cm in diameter and 9 cm high.

Burial: Burial 22/1 found in southwest corner of room, 1.79 m west of east wall and .031 m north of south wall. Infant burial and only skull remains, which faces west. Paint tube found on top of skull and seeds found in vicinity.

Room 23

Dates Worked: July 19–24, 1963.
Form: Rectanguloid.
Dimensions: North wall, 3.85 m; east wall, 5.40 m; south wall, 3.78 m; west wall, 5.25 m. Excavated to depth of 26 cm.

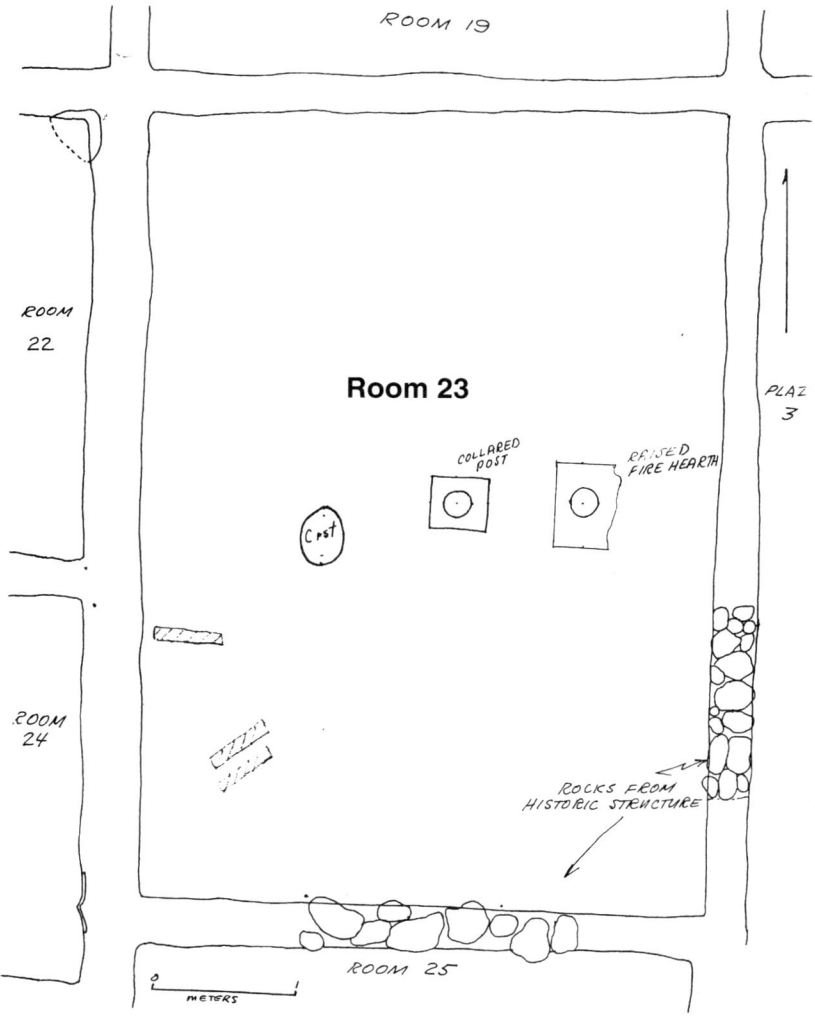

Walls:	Wall height was 62 cm and thickness was 28 cm. Rocks from foundation of historic structure were observed on east and south walls near southeast corner of room. The south wall was about one-half destroyed by the historic construction.
Entrances:	There was a sealed doorway on the north wall (described in Room 19) but any other doorways were likely concealed or destroyed by historic construction.
Floor:	The floor had the typical plastered construction, which was mostly intact but broken by heat in several places.
Roof:	Adobe from roof bonded to floor in several places.
Features:	Raised hearth like that in Room 18 located at 2.75 m north of south wall, 2.88 m east of west wall. The east edge of the hearth was possibly scalloped but it had been damaged. The adobe collar was 4 cm thick and 56 cm wide (north-south). The firepit, 19 cm diameter and 12 cm deep, was filled with ash and soil. An oblong cist, 32 cm (north-south) by 26 cm (east-west) and 51 cm deep, contained two infant skeletons. Found at 2.34 m north of south wall, 1.20 m east of west wall. A collared post, 18 cm in diameter, found at 2.70 m north of south wall, 2.12 m east of west wall. The adobe collar was 37 cm square and 12 cm high.
Burials:	Burial 23/1 found in center of room, 0.27 m west of post collar. Infant burial was in a sitting position facing west. Only artifacts were obsidian flakes and a nodule.

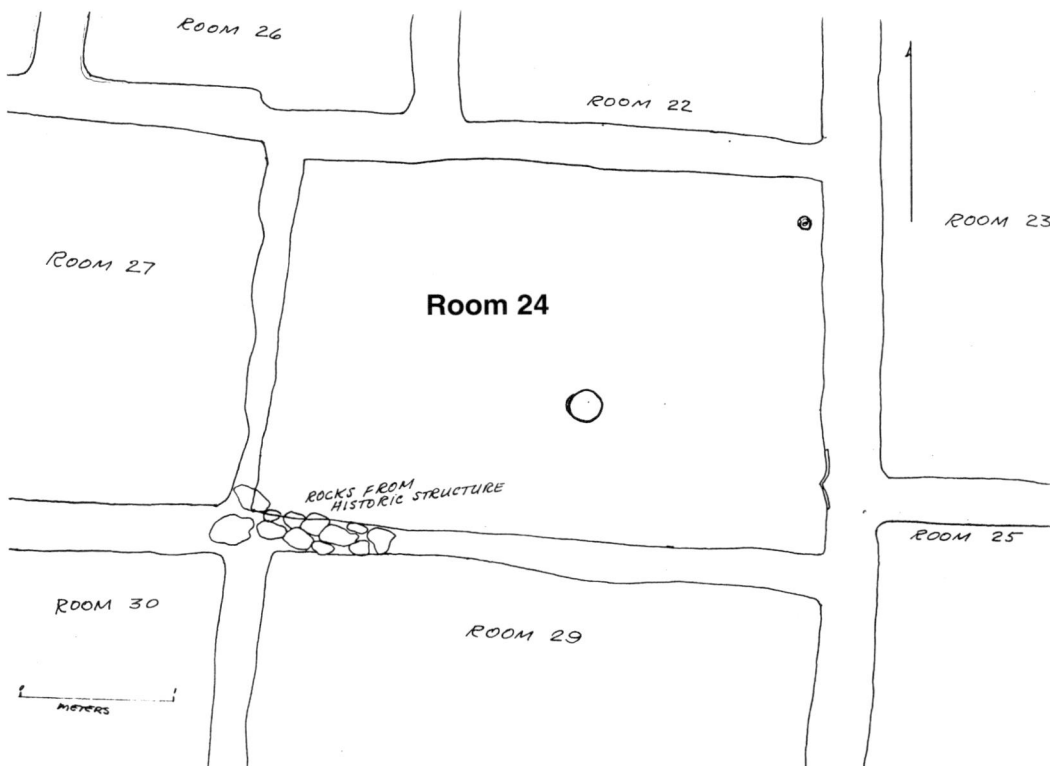

ROOM 24

Dates Worked:	July 22–23, 1963.
Form:	Rectanguloid.
Dimensions:	Not recorded but the depth of excavation was 31 cm.
Walls:	Walls have typical construction and vary in thickness from 24 to 32 cm. The east, south, and west walls have been partially destroyed by the historic structure construction. Plaster is about 4–5 cm thick and is preserved on lower 20–30 cm of wall. Plaster is burned and is an orange or black color.
Entrances:	There was a sealed doorway in the north wall that connects to the south wall of Room 22.
Floor:	Floor is plastered and in excellent condition.
Roof:	Rafters are on the floor with seeds and grass.
Features:	There is a small plastered (burned red) projection on east wall near the floor 52 cm from southeast corner. The feature is 25 cm long, 12 cm high, and 2.5–3.5 cm thick. Use was undetermined.
Burial:	Found in northeast corner of room in a flexed position lying on right side facing north.

ROOM 25

Dates Worked:	July 25–26, 1963.
Form:	Rectanguloid.
Dimensions:	North wall, 3.72 m; east wall, 2.20 m; south wall, 3.60 m; west wall, 2.16 m. Excavated to depth of 34 cm.

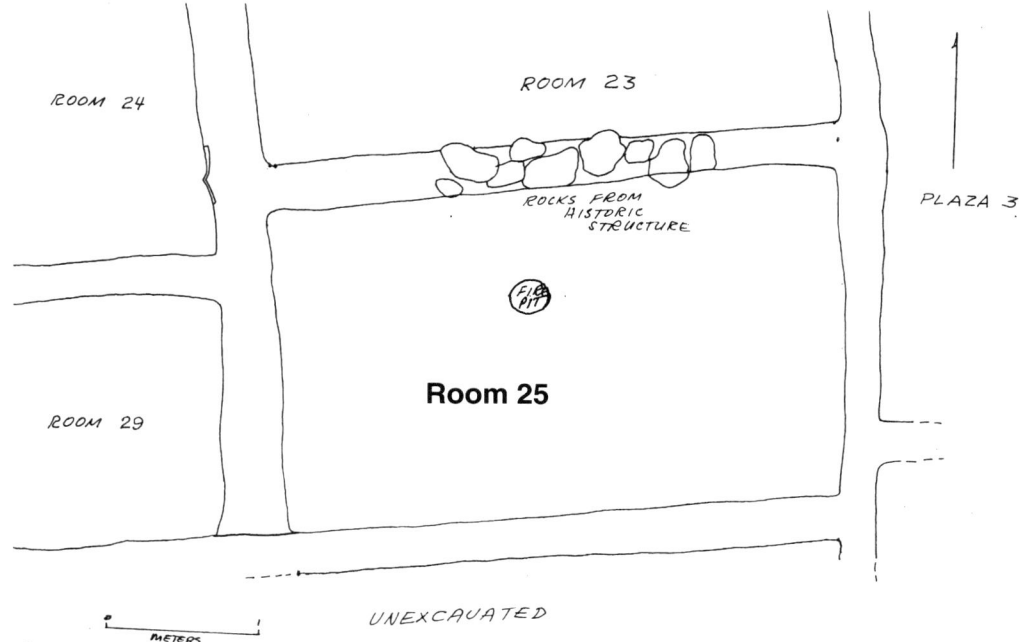

Walls:	Typical construction. Thickness was 24 cm and the height was 81 cm on the north wall and 33 cm on the south. Plaster is missing over large areas of the room and there is evidence of root and rodent activity. The west wall abuts into the south wall.
Entrances:	None visible.
Floor:	Plaster in excellent condition.

ROOM 26

Dates Worked:	July 27–30, 1963.
Form:	Rectanguloid.
Dimensions:	North wall, 2.82 m; east wall, 2.97 m; south wall, 2.15 m; west wall, 2.66 m.
Walls:	Typical construction with a thickness of about 23 cm. The average height of the walls is 79 cm with the north wall only 48 cm high. No plaster is visible. There is a rock imbedded in base of west wall (center). West wall has a joint (1.50 m from south wall) that may indicate an ad-

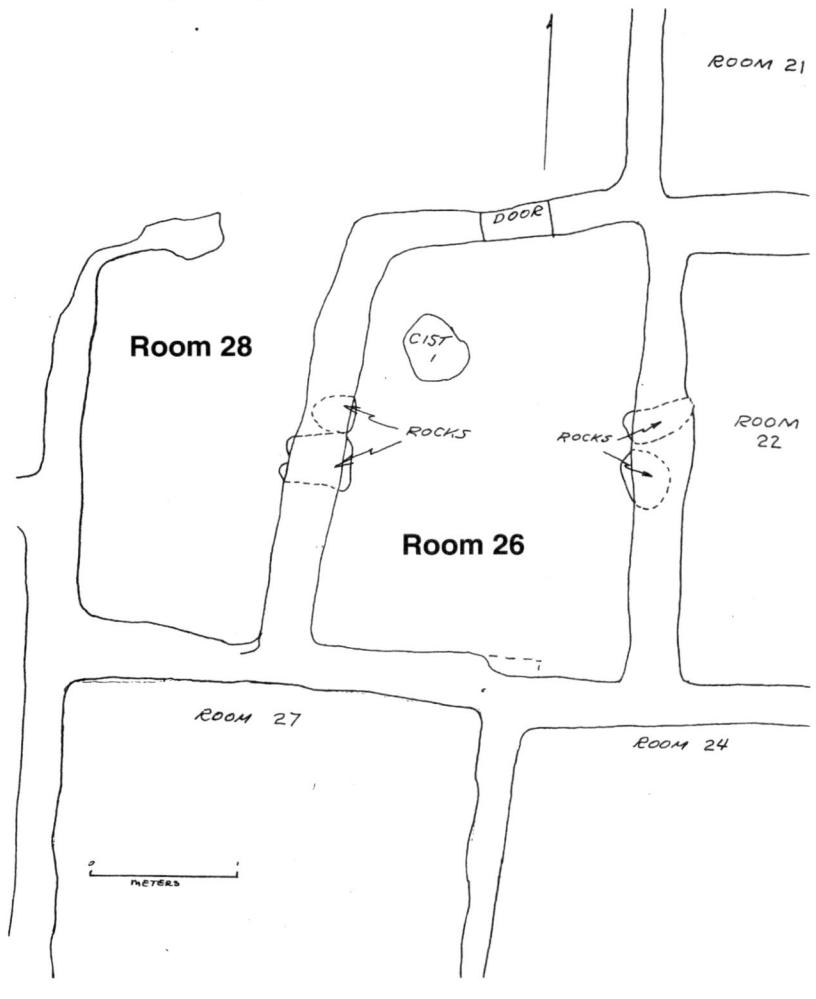

Entrances:

dition to northwest corner of the room at a later time. North wall of the settlers' structure passes through the east and west walls.

There is a sealed door in north wall (59 cm from east wall). Door (53 cm wide) is flush with floor and sealed with large rocks, adobe, and a large skull—possibly bison. No plaster was found on either side and it could not be determined from which side the door was sealed.

Floor:

The floor is packed earth with stones and is uneven and rough and it may have been plastered.

Features:

There is a cist in the floor, 56 cm deep, 26 cm east-west, and 32 cm north-south. It is located 42 cm from west wall and 63 cm from north wall. Fill of the cist is soft and a half shell bracelet and sherds were found within. In fill, seven manos were found in northwest corner of the room heaped in a pile. A mica pendant also found. On the floor there was a quartz crystal, asbestos fibers, a stalactite, yellow clay, and an abundance of bone, both on the floor and in the fill.

ROOM 27

Dates Worked: July 27, 1963.
Form: Rectanguloid.
Dimensions: North wall, 2.83 m; east wall, 2.45 m; south wall, 2.62 m; west wall, 2.57 m. Excavated to depth of 25 cm.

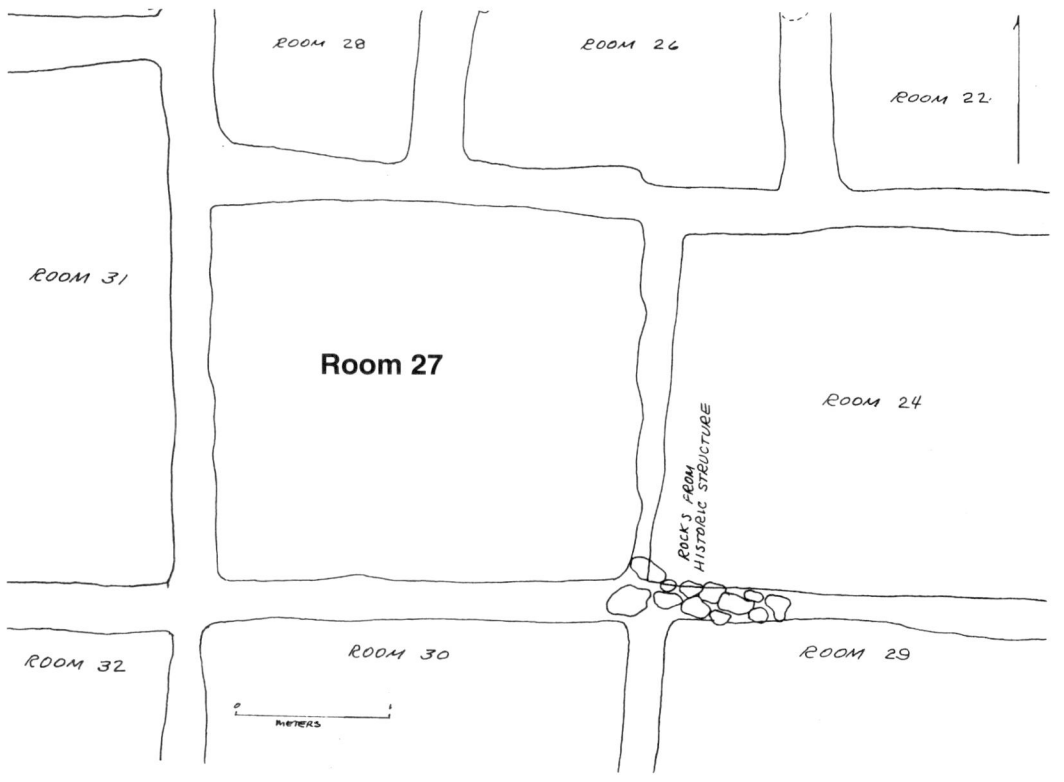

Walls:	Typical construction but the south wall is partially destroyed by historic construction. Plaster is brittle, burned orange and black and extends up the wall to a height from 63–75 cm. Some small boulders were embedded in the south wall.
Entrances:	None visible.
Floor:	Floor was plastered and in excellent condition.
Features:	There were the remains of a possible corner hearth in the northeast corner. It was 34 cm high, 20 cm wide, and 4 cm deep.

ROOM 28

Dates Worked:	July 27, 1963.
Form:	Rectanguloid.
Dimensions:	North wall, 2.69 m; east wall, 2.77 m; south wall, 1.37 m; west wall, 2.35 m. Excavated to depth of 12 cm.
Walls:	Typical wall construction, 53 cm high and 24 cm thick.
Entrances:	In the northeast corner is a open doorway, 82 cm wide and 50 cm high, at the floor level that opens into the plaza.
Floor:	Floor is packed earth and is irregular and poorly defined.
Features:	There is a circular fire pit, 20 cm in diameter, 11 cm deep, and located at 1.98 north of south wall and 1.14 west of east wall. The fire pit is damaged and a portion is missing. There is an ash layer, 3 cm below the floor, that may be an original plaza surface. The small size of the room and dirt floor may indicate use as a storage space or a pen. Fire pit, however, indicates human occupation and possible use as a workshop adjoining the plaza.

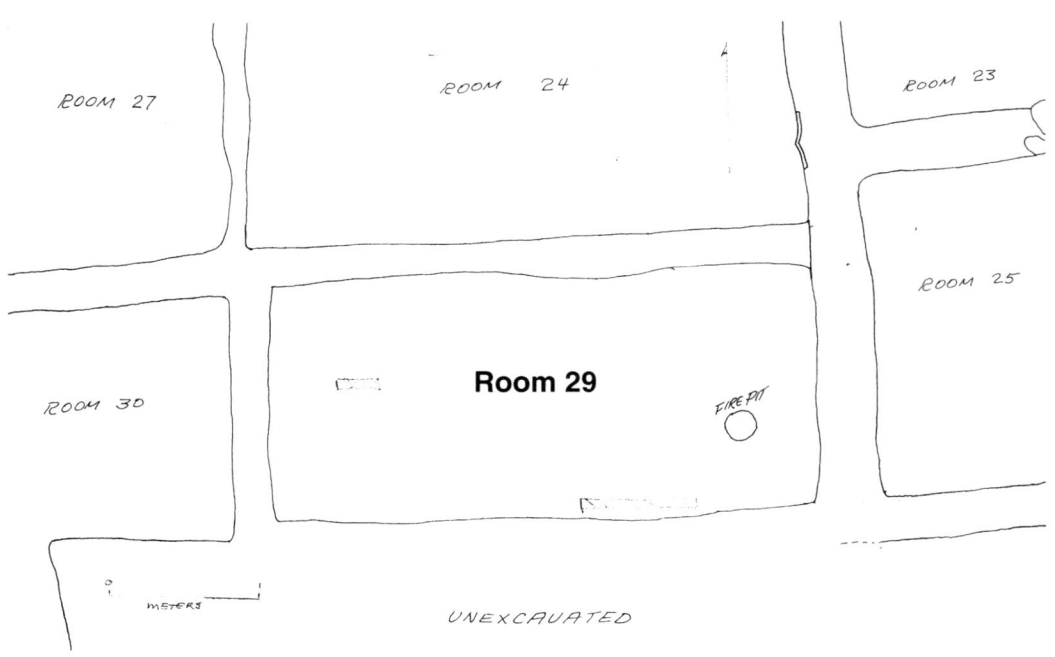

ROOM 29

Dates Worked:	July 7 and August 5, 1963.
Form:	Rectanguloid.
Dimensions:	North wall, 3.55 m; south wall, 3.58 m; west wall, 1.58 m; east wall 1.60 m. Depth of excavation was 22 cm.
Walls:	Walls had typical construction and reach a height of 60 cm except on the south wall, which is only 49 cm high. Thickness of the walls is 26 cm and some root action is visible. The plaster is poorly preserved except from 10–20 cm above the floor and lower. Some rocks are imbedded in the wall near the floor level. The north wall abuts into the east wall.
Entrances:	None visible.
Features:	A poorly preserved circular fire pit was found at 31 cm west of east wall and 57 cm north of south wall. It was 21 cm in diameter and 10 cm deep.
Burial:	An infant burial found in northwest corner of room, lying on back with head toward west and facing up.

ROOM 30

Dates Worked:	July 30–31, 1963.
Form:	L-shaped.
Dimensions:	North wall, 2.76 m; south wall, 1.55 m and 1.25 m; west wall, 3.29 m; east wall, 1.58 m and 0.80 m.
Walls:	Same construction style as all rooms and the height of the walls is 78 cm

on the north side and 60 cm on the south. The walls are 22 cm thick and three large rocks are embedded in the wall near the floor on the north wall. Plaster is preserved up to 40 cm above the floor and on north wall brush marks are visible.

Entrances: None visible.

Floor: The floor is in poor condition owing to root action. The plaster is largely removed, pitted, and uneven.

Roof: Two burned rafter fragments were found on the floor with decomposed reeds found in the form of impressions.

Features: An ash layer was found 17 cm below the floor.

ROOM 31

Dates Worked: July 30–August 2, 1963.

Form: Rectanguloid.

Dimensions: North wall, 4.66 m; south wall, 4.15 m; west wall, 3.48 m; east wall, 3.53 m. Depth of excavation was 18–30 cm below contemporary surface to top of walls.

Walls: Walls had typical construction and were 76 cm high and 25 cm thick. Wall is deeply eroded in the northwest corner. Plaster is preserved well and comes within 15 cm of the tops of the walls. Plaster is 4 mm thick and three or four layers are evident.

Entrances: There is a keyhole-shaped doorway in the center of the south wall

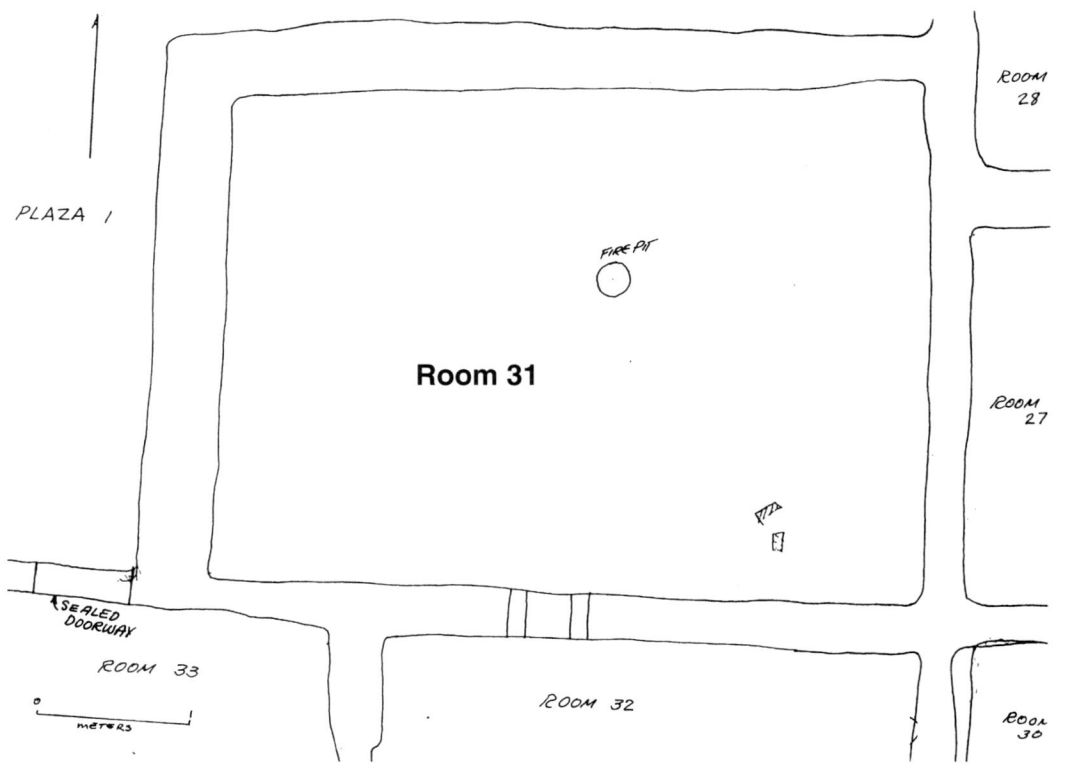

(lower left corner of door is 2.12 m from southeast corner of room). It is sealed with mud from Room 32. Doorway is sealed poorly in that it is uneven and not flush with wall. Dimensions are 49 cm on top and 20 cm on base of doorway.

Floor: There is good evidence of multiple plastering and there is some spalling of the plaster around the fire pit.

Features: There was a fire pit located at 2.62 m north of south wall and 2.24 m west of east wall. The depth was 16 cm and it is filled with ash mixed with soil. The interior is plastered but there is not a prominent lip. There was some spalling of the plaster around the edges. A posthole was found at 2.73 m north of south wall and 2.27 m west of east wall. The diameter is 24 cm and the interior has been roughly plastered but no lip was formed. Two manos and a bone awl were found.

ROOM 32

Dates Worked: August 5–6, 1963.
Form: Almost square.
Dimensions: North wall, 3.50 m; south wall, 2.47 m; west wall, 2.08 m; east wall, 1.29 m. Depth of excavation was 31 cm to top of wall.
Walls: Walls have typical construction and the thickness is 30 cm; the east wall is 70 cm high. Walls are burned from the top to the floor and are colored from orange to black. Burning preserved the walls and plaster. The south half of the west wall had been destroyed by historic construction. East wall is unusual in that it appears to have been constructed in one

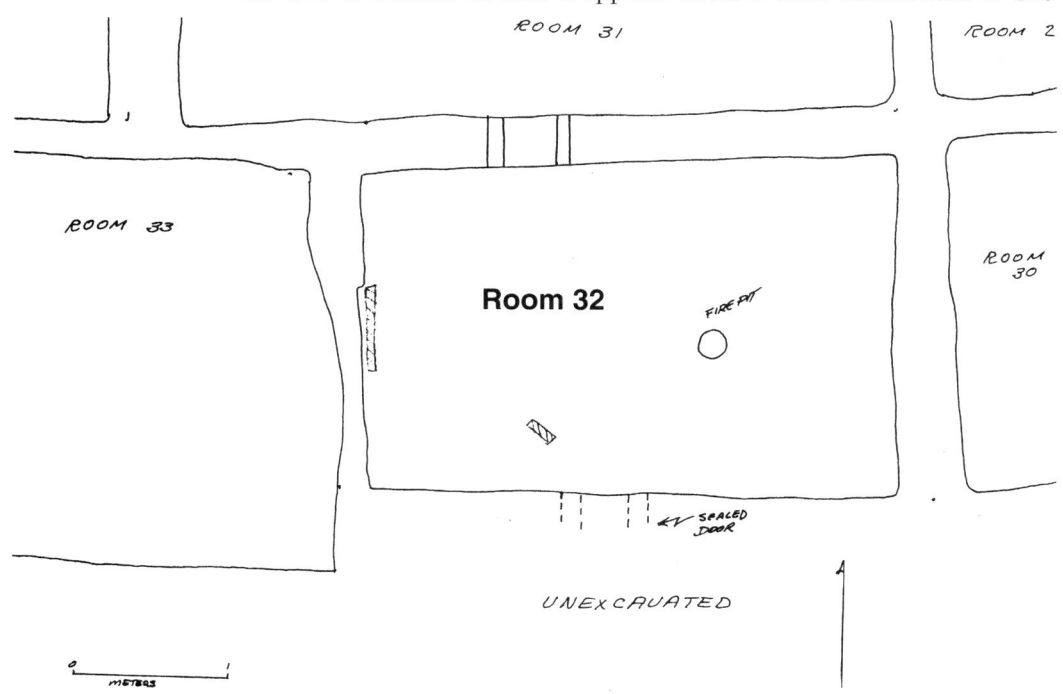

course 53 cm high. Top of the course is smooth and another course, 22 cm thick, was added.

Entrances: There is a sealed door in center of north wall that connects to Room 31. On the south wall there is a possible sealed keyhole door (lower left corner is 1.99 m from southeast corner of room). The outline is obscured by very hard plaster and the dimensions could not be revealed by removing the plaster.

Floor: The plaster is buckled and broken because of the extreme heat. It is especially charred and baked in the western two-thirds of the room.

Roof: The roof was approximately 25 cm thick and most of the adobe was completely oxidized. Fired roof plaster, burned reeds, grass, and fragments of rafters were located.

Features: There was a circular fire pit (96 cm north and 3.30 m west) that was 20 cm in diameter and 12 cm deep, with a 12 cm collar. Fire pit was damaged on the north side. Corn, beans, and a very small amount of bone were also found in the room.

ROOM 33

Dates Worked: August 5–6, 1963.
Form: Rectanguloid.
Dimensions: North wall, 4.66 m; south wall, 4.15 m; west wall, 3.48 m; east wall, 3.53 m. Depth of excavation was 18–30 cm.

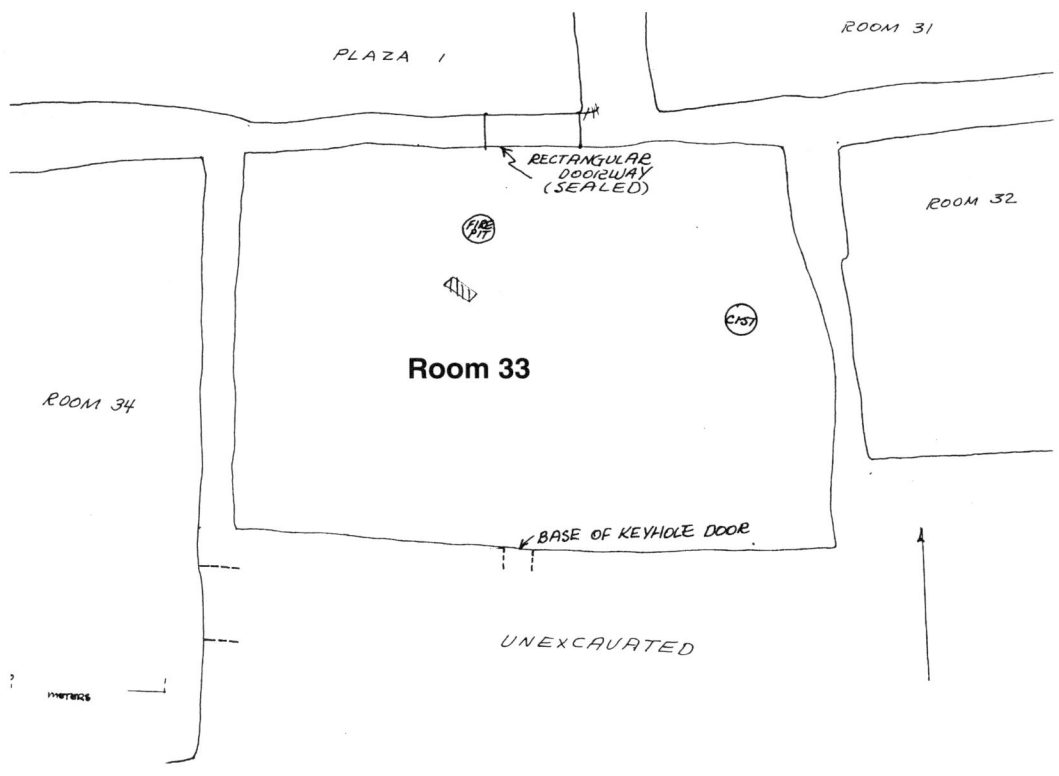

Walls:	Walls reached height of 59 cm and were 24 cm thick. Plaster was preserved to the top of the remaining walls. Root and rodent action was visible in the walls.
Entrances:	On the north wall (1.26 m from northeast corner of room) is a 60-cm-wide rectangular door that was flush with the floor. Door is sealed and filled with adobe. On the south wall the base of the keyhole door was found. The lower left corner is 1.76 m from southwest corner of the room. The narrow portion of the doorway is 20 cm on top and 15 cm on bottom and is 16 cm above the floor.
Floor:	Floor was plastered in good condition but was irregular in places.
Features:	A circular fire pit was located at 0.73 m north of south wall and 2.07 m west of east wall. A storage pit was located at 52 cm west of east wall and 1.55 m north of south wall. Dimensions of the pit are 38 cm east-west, 40 cm north-south, with a depth of 42 cm and a diameter of 19 cm. There was a historic excavation of the room. A circular hole, approximately 3 m in diameter, was dug into the room and about two-thirds of the east wall was removed. Historic artifacts (glass, stove fragment) were found in the fill.

ROOM 34

Dates Worked:	August 5–6, 1963.
Form:	Rectanguloid.
Dimensions:	North wall, 3.49 m; south wall, 2.96 m; west wall, 4.16 m; east wall, 4.05 m. Depth of excavation was 28–41 cm from the contemporary surface to the tops of the walls.
Walls:	Height of the east wall was 60 cm (22 cm thick) but historic excavation has removed most of the north and west walls and a portion of the south wall. Plaster was preserved to 21 cm on north wall and was burned to an orange and black color. Some root and rodent action was visible.
Entrances:	The east wall had a rectangular doorway that leads to an unexcavated room. Lower right corner of the doorway is 80 cm from southeast corner of the room. Door was sealed from Room 34 side. The doorway began about 23 cm above the floor and the opening was 49 cm wide and 37 cm high. The south wall had a possible door in the center of the wall but it was very poorly defined. It was possibly rectangular and the right corner of the door was 1.14 m from the southwest corner of the room. The approximate measurements were 46 cm wide and the opening began 20 cm from the floor.
Floor:	Plaster was largely removed, probably by heat. Surface was uneven and pitted.
Features:	There was a circular fire pit located 2.0 m north of south wall and 2.28 m west of east wall. It was 23 cm east-west and 19 cm north-south, and 19 cm deep. It had a plastered collar (9 cm wide on the south side) that was partly destroyed. A probable ash lip also was found on the west side.

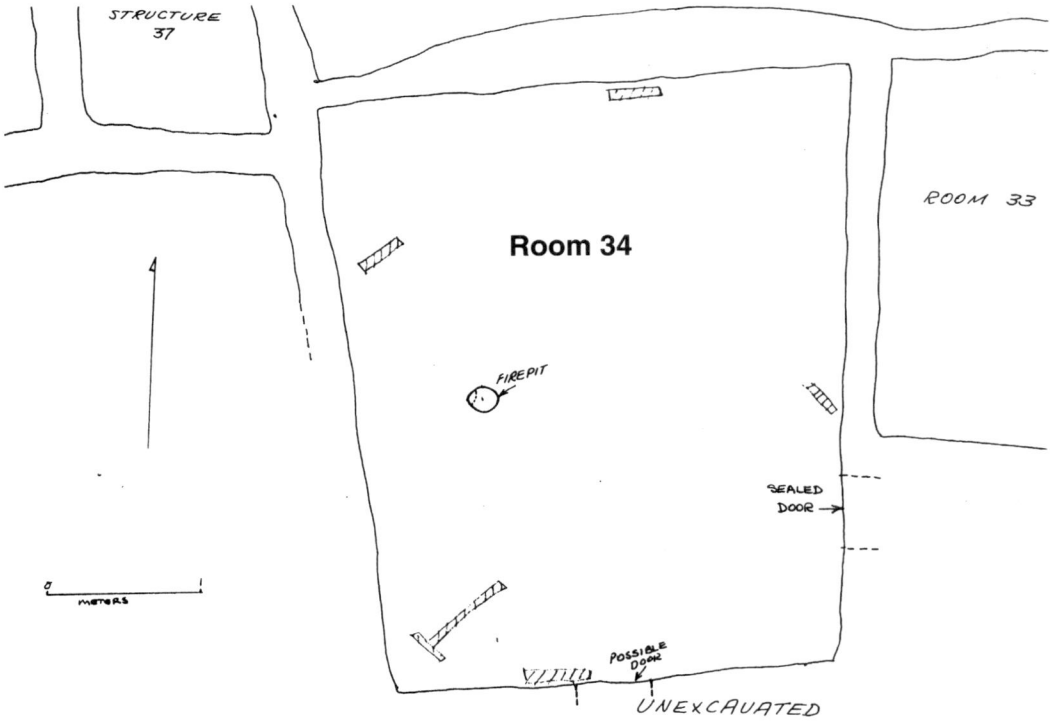

Burials: Burial 34/1 was located in the southeast corner, 0.048 m north of south wall and 0.28 m west of east wall. The burial was lying on its left side, head was to the north and facing east. Artifacts include shell beads, two turquoise pendants, two shell pendants, one turquoise pendant blank, one paint tube, and one concretion with a groove. Burial cist was plugged with a 15-cm-thick packed adobe layer, and hole was dug with a flange to accommodate the plug.

Burial 34/2 was against the east wall and was lying on its left side with the head to the south, facing west. Corn, seeds, leather, and berries were found in association with the burial.

Room 35

Dates Worked: August 7–8, 1963.

Form: Rectanguloid.

Dimensions: North wall, 2.22 m; south wall, 2.10 m; west wall, 3.16 m; east wall, 3.62 m. Depth of excavation was 20 cm from the contemporary surface to the tops of the walls.

Walls: Walls were 48 cm high and 26 cm thick. Plaster (3 cm thick) was burned orange and black and is preserved to about 36 cm above floor. West one-half of north wall has been destroyed by historic excavation. There was a short dogleg in south wall for no apparent reason. Plaster here was 7 cm thick.

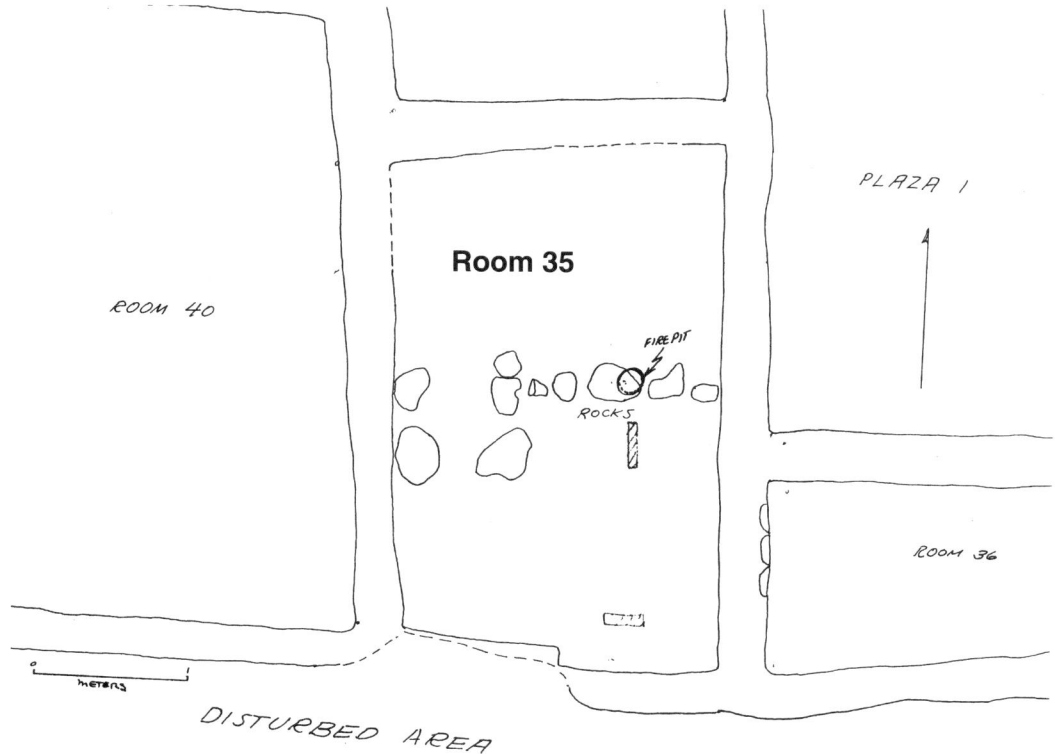

Entrances:	None visible but possibly destroyed.
Floor:	The floor was plastered but much of it was destroyed by heat, pitted and irregular.
Features:	There was a circular fire pit located 0.71 m west of east wall and 1.88 m north of south wall (diameter is 19 cm and depth is 12 cm). A collar (11 cm wide) was preserved on the south and west sides only, and an ash lip was found on the south side under a large stone. Large stones on floor were oriented in an east-west direction and appeared to divide the room. A burial was found in the southwest corner. The room is situated on the north edge of a disturbed area.
Burial:	Burial 35/1 was found in the southwest corner of the room, 2 cm above the floor surface. Burial was in the room fill. Burial is on left side with the head to the south, facing west. There were no associated artifacts.

ROOM 36

Dates Worked:	August 6, 1963.
Form:	Rectanguloid.
Dimensions:	North wall, 2.20 m; east wall, 1.26 m; south wall, 2.18 m; west wall, 1.15 m. Depth of excavation was 19 cm from the contemporary surface to the tops of the walls.
Walls:	Height of the walls was 41 cm and the thickness was 27 cm. Three rocks

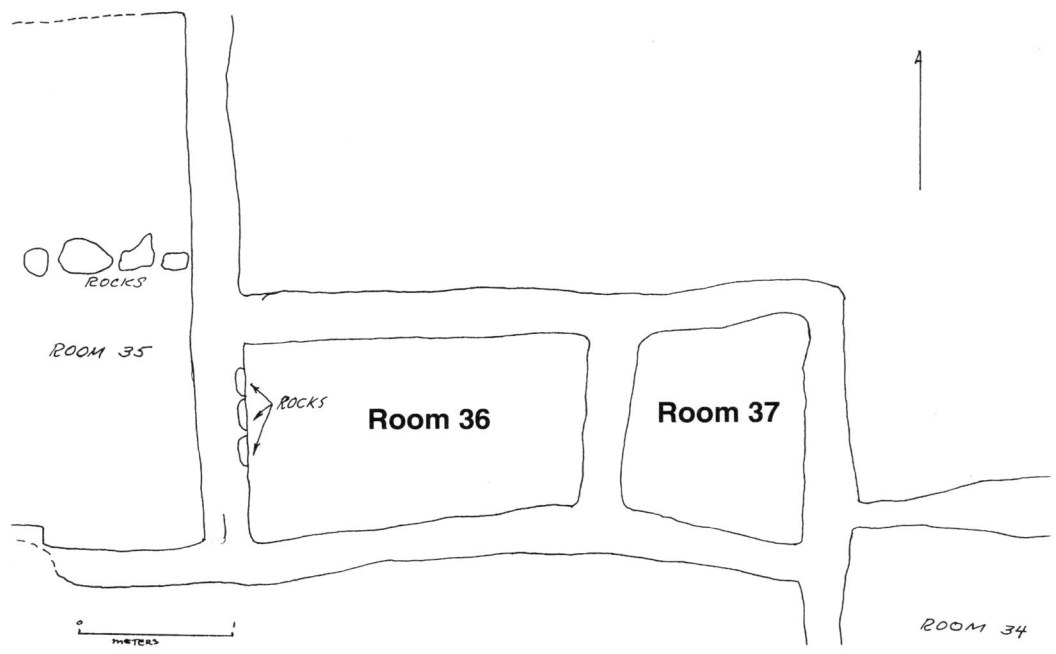

(approximately 16 cm by 16 cm in size) were found embedded in the west wall near the floor. No plaster was found.

Entrances:	No visible doorways.
Floor:	The floor was compacted earth and was irregular.
Features:	This is a possible storage room.

ROOM 37

Dates Worked:	August 8, 1963.
Form:	Rectanguloid.
Dimensions:	North wall, 1.04 m; east wall, 1.40 m; south wall, 1.17 m; west wall, 1.28 m. Depth of excavation was 16 cm from the contemporary surface to the tops of the walls.
Walls:	Walls were 42 cm high and 26 cm thick. Root and rodent damage visible on walls and floor. There was no plaster on the walls.
Entrances:	None visible.
Floor:	The floor was compacted and was irregular.
Features:	This is a small structure possibly used for storage.

ROOM 38

Dates Worked:	August 7–8, 1963.
Form:	Almost square.
Dimensions:	North wall, 2.23 m; east wall, 1.98 m; south wall, 2.16 m; west wall, 2.0 m. Excavated to 20 cm at north and 32 cm at west side.
Walls:	Wall height was 52 cm but was eroded to 35 cm at south end of east

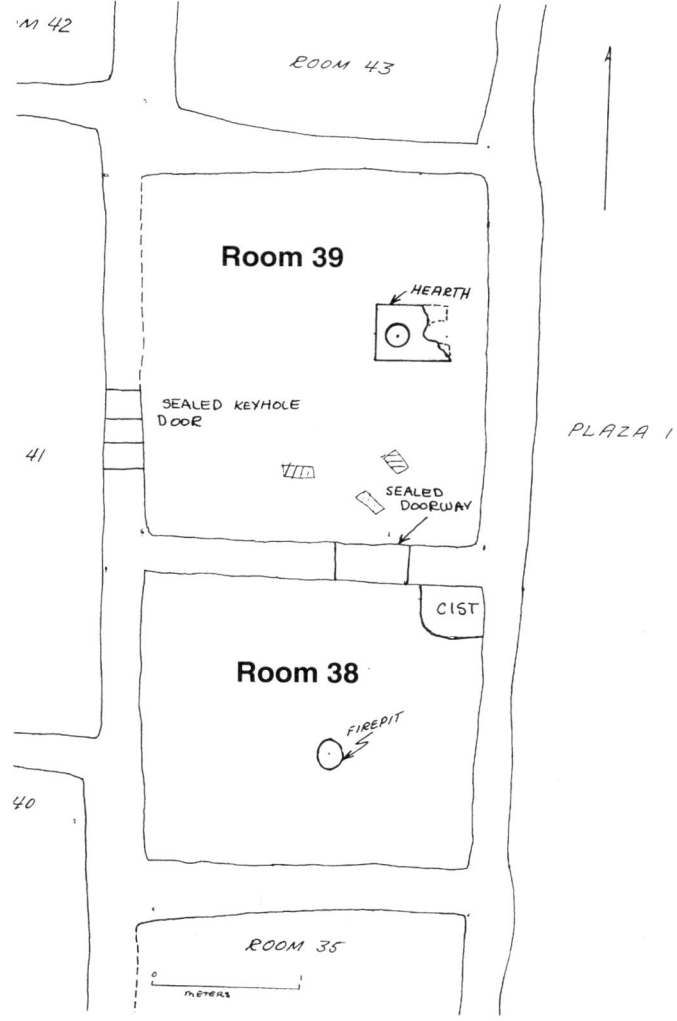

wall. Wall thickness was 28 cm. Historic excavation removed south wall to 6 cm above floor. Rocks were found embedded in exterior of east wall at base. Plaster was absent except on west wall where it was preserved to 43 cm above floor and burned to orange and black color.

Entrances:
A wedge-shaped doorway, sealed from Room 38, was found in the north wall. Top of opening was 46 cm, which curved on west side to 15 cm at floor. There was only a slight taper on east side. Lower left corner was 1.30 m from northeast corner of room.

Floor:
Plaster was damaged by heat. Pitted sections over most of surface reveal pebbles in floor.

Roof:
A few pieces of charcoal in fill at floor level.

Features:
Oval firepit, 23 cm (north-south) by 15 cm (east-west) and 14 cm deep, found at 0.74 m north of south wall, 1.06 m east. A cist, 34 cm (north-south) by 38 cm (east-west) and 35 cm deep, was also located in the northeast corner of the room.

ROOM 39

Dates Worked:	August 7–9, 1963.
Form:	Almost square.
Dimensions:	North wall, 2.30 m; east wall, 2.49 m; south wall, 2.20 m; west wall, 2.44 m. Depth of excavation 13 cm.
Walls:	Wall height was 60 cm but eroded to 39 cm on east side. Wall thickness was 28 cm. Historic excavation removed north half of west wall and west half of north wall. Walls speckled with white mineral, possibly calcium carbonate. Only small patch of 5-cm-thick plaster in room, on west wall to 25 cm above floor. Burned orange and black, flaked off in sheets of 1–2 mm thick.
Entrances:	A sealed doorway was located in south wall, which was common to Room 38, north wall. A T-shaped doorway in west wall was sealed from Room 41 with adobe. Upper opening was approximately 48 cm wide, lower opening 21 cm wide and 31 cm high; sill was 19 cm above floor. The lower left corner was 0.45 m from southeast corner of room.
Floor:	Plaster was poorly preserved. Pitted and damaged by rodent activity.
Features:	Rectangular raised hearth was located at 1.37 m north of south wall, 1.67 east of west wall. The hearth, damaged on east side, had a 3-cm-thick adobe collar that was 35 cm (north-south) on intact side, and 33 cm (east-west) on damaged side. The circular firepit, 16 cm in diameter and 15 cm deep, was filled with ashes and burned wood. Ash removal lip was on east side.

ROOM 40

Dates Worked:	August 8–9, 1963.
Form:	Rectanguloid.
Dimensions:	North wall, 2.29 m; east wall, 4.16 m; south wall, 2.50 m; west wall, 4.22 m. Excavated to 10–20 cm depth.
Walls:	Wall height was 79 cm and thickness 22 cm. The room was near disturbed area. East wall partially destroyed by historic excavation and west wall was indistinct. Sections of wall destroyed by root and rodent action. Plaster preserved to 33 cm above floor, east wall.
Entrances:	Rectangular doorway in north wall was sealed from Room 41. The opening was 46 cm wide and the sill was 21 cm above floor. The lower right corner 85 cm from northeast corner of room.
Floor:	Plaster and floor in poor condition. Most of surface was absent.
Features:	A circular firepit, 14 cm diameter and about 8 cm deep, was located at 2.14 m north of south wall, 1.69 m east of west wall. Slight elevation of floor around firepit indicative of absent raised hearth. Rectangular hearth, 23 cm by 23-by-17 cm high, set on diagonal into southeast corner of room.

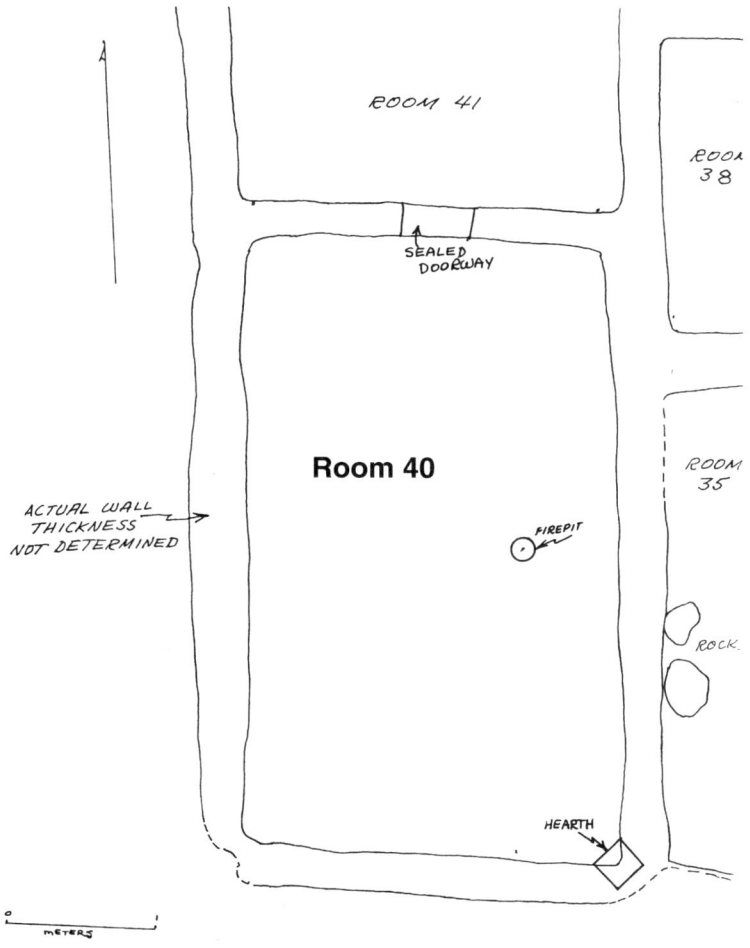

ROOM 41

Dates Worked:	August 9, 1963.
Form:	Rectanguloid.
Dimensions:	North wall, 2.59 m; east wall, 4.11 m; south wall, 2.52 m; west wall, 4.01 m. Depth of excavation was 19 cm from the contemporary surface to the tops of the walls.
Walls:	Wall height was 61 cm and thickness was 22 cm. Plaster was preserved to top of walls on east wall and found sporadically on other walls. Vertical cracks observed in north wall. Historic excavation destroyed part of the south wall. Rodent and root action was observed but no wall abutting was visible.
Entrances:	The east wall had a keyhole doorway that was sealed with adobe from this room. Lower left corner of doorway was 1.77 m from northeast corner of the room. The doorway began 19 cm off the floor and was 48 cm on top and 21 cm on the bottom. Two other doorways are in the room. The south wall has a sealed door that connects to Room 40 and

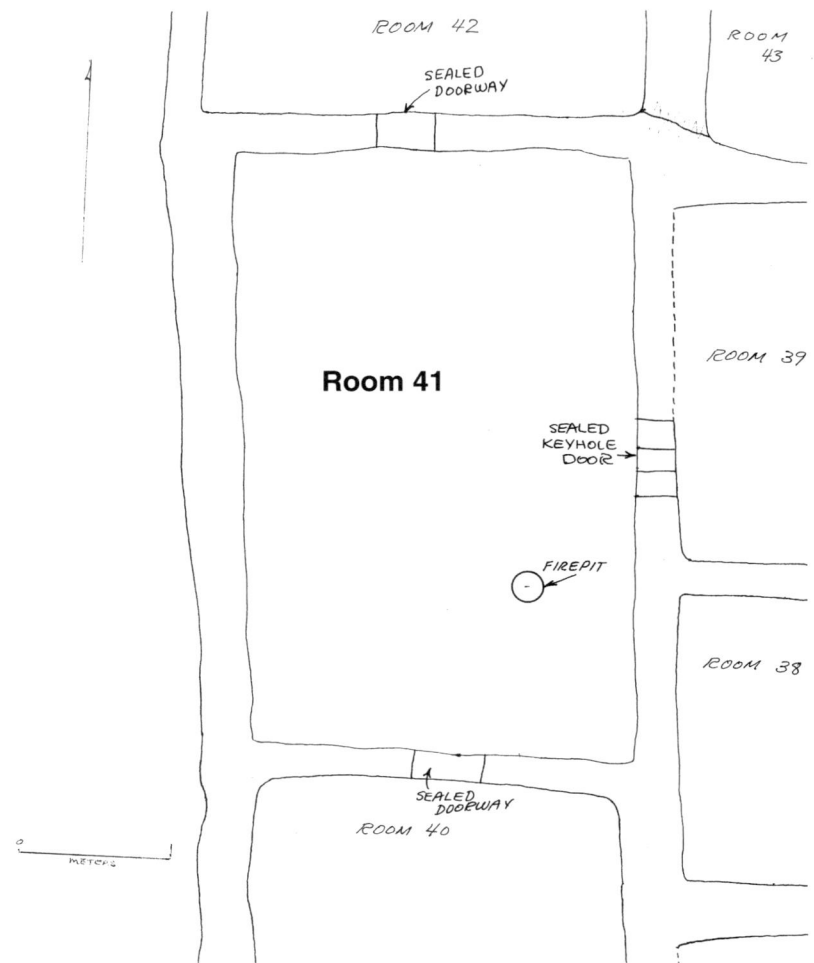

the north wall has a sealed door that connects to Room 42.

Floors: The floor may have been plastered. Most of the plaster is missing and the roof adobe is bonded with the floor.

Features: A circular fire pit was located at 1.12 m north of south wall and 0.75 m west of east wall. The pit was 20 cm in diameter and depth was unknown because a large root grew through the feature.

ROOM 42

Dates Worked: August 13–14, 1963.
Form: Rectanguloid.
Dimensions: North wall, 2.82 m; east wall, 3.37 m; south wall, 2.90 m; west wall, 3.52 m. Depth of excavation was 28 cm from the contemporary surface to the tops of the walls.
Walls: The wall height was 38 cm on south and 34 cm on east, and thickness was 22 cm. Plaster was preserved to about 34 cm above the floor. Historic excavation had partly destroyed east wall. There was some evi-

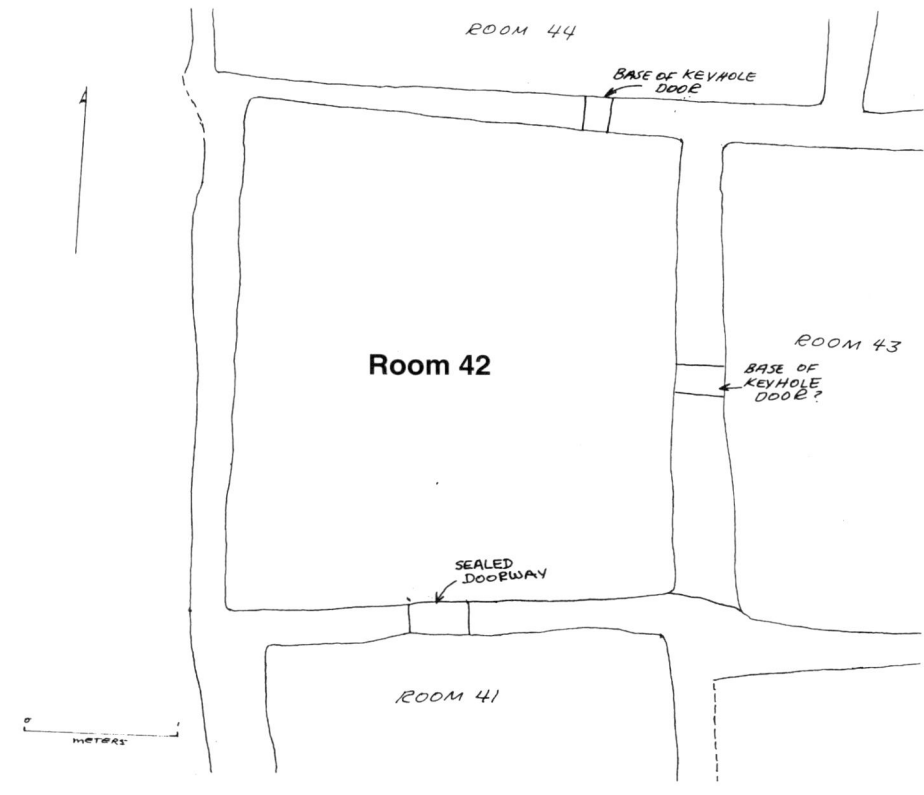

dence of root and rodent damage. The east wall abuts into the south wall.

Entrances: South wall has a rectangular doorway that was sealed from both sides. The lower left corner of the doorway was 1.14 m from the southeast corner of the room. The doorway base was 10 cm from the floor and the feature was 42 cm wide. The north wall had a sealed T-shaped door that connects to Room 44 and the east wall has a possible key-shaped door that connects to Room 43.

Floor: The floor is plastered and is irregular but smooth.

ROOM 43

Dates Worked: August 9–10, 1963.
Form: Rectanguloid.
Dimensions: North wall, 2.28 m; east wall, 3.28 m; south wall, 2.04 m; west wall, 3.21 m. Depth of excavation was 17 cm from the contemporary surface to the tops of the walls.
Walls: Height was 48 cm and thickness 28 cm. Historic excavation had removed most of the west and south walls and erosion and root action had lowered the north and east walls. Plaster was on found on a small patch on east wall 16 cm above the floor.
Entrances: There was a possible keyhole door on the west wall. Lower right corner

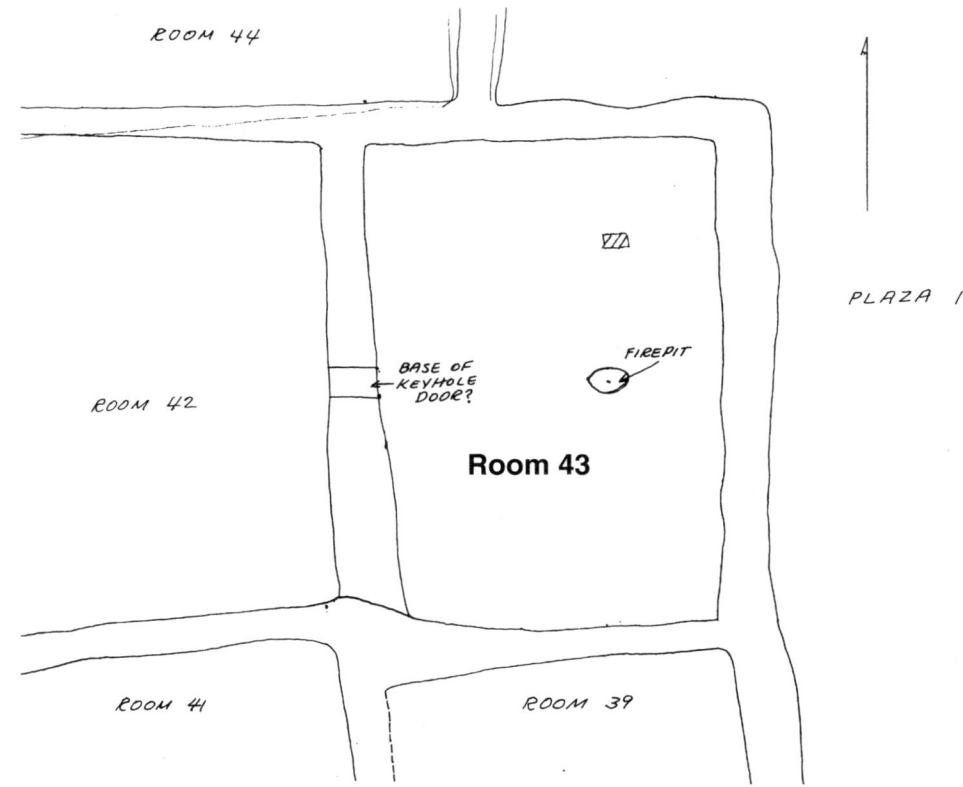

was 1.52 m from northwest corner. Width of the lower part of the door was 19 cm.

Floor: Floor was plastered but uneven and disturbed by roots.

Features: An oval fire pit was found at 1.69 m north of south wall and 0.62 m west of east wall. It was 25 cm east-west and 21 cm north-south and had a depth of 14 cm. There was an ash lip to the east. A possible floor polisher was also found.

ROOM 44

Dates Worked: August 13–15, 1963.

Form: Almost square.

Dimensions: North wall, 3.80 m; east wall, 3.88 m; south wall, 3.88 m; west wall, 3.80 m. Excavated to a depth of 15 cm.

Walls: Wall height was 38 cm; thickness was 24 cm at east and north, 18 cm at south. West wall almost completely destroyed by root action, many briar roots in other walls. In northeast corner, rock embedded in wall over 8-by-16-cm area. This room added to adjacent rooms as east and west walls abut south wall, common to Rooms 42 and 43. Plaster, 4 cm thick, preserved to top of south wall, burned to black and orange color, and very brittle.

Entrances: A sealed T-shaped doorway in north wall was common to Room 45's

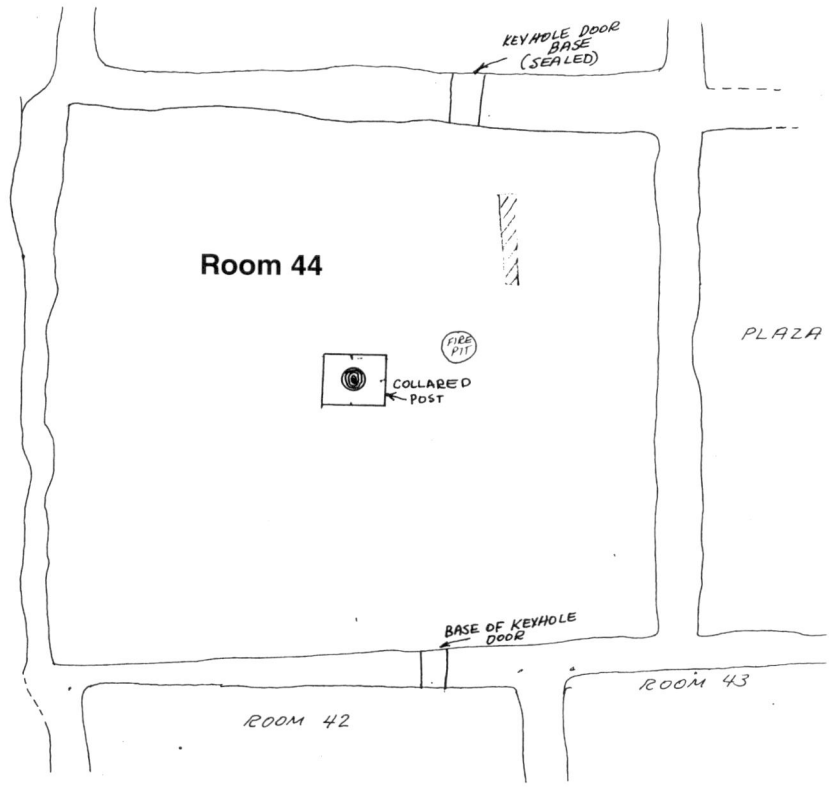

south wall. In south wall, 16-cm-wide base of a T-shaped door sealed with adobe from both sides. Lower left corner was 1.42 m from southeast corner of room.

Floor:
The floor had the typical plastered construction, but was very irregular and uneven. Burned to dark gray color, buckled and broken.

Features:
A circular firepit, 20 cm diameter and 13 cm deep, was found at 1.99 m north of south wall and 2.55 m east of west. The ash lip was oriented toward east. Collared post, with charred 16-cm-diameter post in situ, was located at 1.93 m north of south wall and 1.07 m east of west wall. It had a rectangular adobe collar, 6 cm thick, 38 cm (north-south), and 36 cm (east-west). Top of fragmented post was 21 cm above collar.

Burial:
Burial 44/1 was located on the north side of the room, 1.40 m west of east wall and 3.43 m north of south wall. Burial was disintegrated except for one tooth.

ROOM 45

Dates Worked:
August 15–16, 1963.

Form:
Almost square.

Dimensions:
North wall, 3.43 m; east wall, 3.45 m; south wall, 3.72 m; west wall, 3.49 m. Excavated to 17 cm depth.

Walls:
Walls had typical construction, but were eroded on all sides to 23 cm

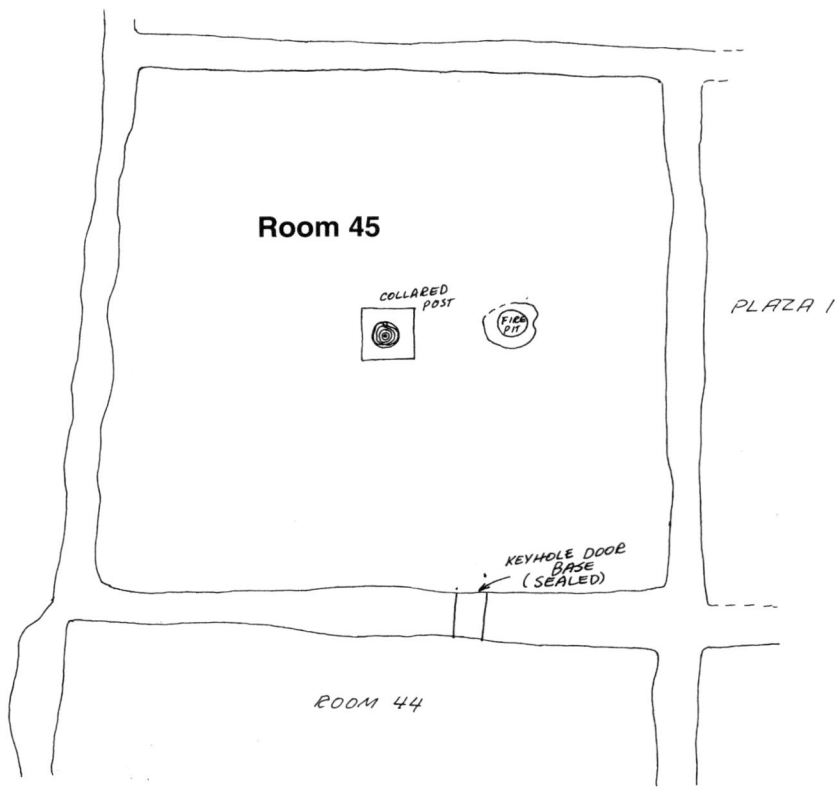

height. Thickness was 24 cm. South half of west wall destroyed by root action. Plaster remains on south wall to 16 cm above floor, east wall to 23 cm. Plaster burned to an orange color.

Entrances: Base of T-shaped doorway, 18 cm wide, was located in south wall, flush with floor. The doorway was sealed from both sides. The lower left corner of doorway was 1.23 m from southeast corner of room.

Floor: The floor had the typical construction, but was irregular in areas.

Features: A collared oval firepit, 19 cm north-south, 21 cm east-west, and 12 cm deep, found at 1.79 m north of south wall and 2.55 m east of west wall. Collar was 37 cm in diameter with an ash lip to east. A collared post was found at 1.73 north of south wall, 1.76 east of west wall, with a 17 cm post burned flush with top of collar. The square collar, 36 cm per side and 16 cm high, was damaged on top by root action.

THREE

Plant Materials from the 1963 Excavation

Hugh C. Cutler and William C. Cutler

Cultivated plants should be considered artifacts because they are to a large extent the creations of man, are manipulated and changed by his actions, and seldom can exist without him. They are like pottery in many respects. Corn is a good example. Most American Indians grew many different kinds, some of them for special uses, just as different kinds of pottery were made for distinct purposes. For a thorough study of the corn of any habitation, it is essential to distinguish the various elements and then to compare them with similar elements from other sites. This is rarely possible because variants of cultivated plants can seldom be identified from archaeological remains.

Color and composition of the grains of corn were and still are used by many Indian groups to identify their varieties. These characteristics were destroyed by burning, which carbonized vegetal remains of some open sites, and made this usually perishable material resistant to decay. Fortunately, there are some cob characteristics that can be used to distinguish clusters recognized by the original inhabitants of a site and, frequently, by modern Indians. Such rough groupings were discovered in studies on corn from the Clanton Draw site (29 HI SAR 62-1) and Box Canyon site (29 HI SAR 62-2) (McCluney 1965a), and similar groupings can be detected in the present sample of plant materials from the Joyce Well site.

Practically all of the material from this site was carbonized at moderate temperatures and the estimated shrinkage in the cupule width is about 15 to 20 percent. Most of the cobs from Room 32, room fill (cat. no. 2337), had a metallic sheen and were hard and brittle, characteristics that we have been able to reproduce by heating cobs at a relatively higher temperature than is required to duplicate the other specimens

from this site. The greater heat results in more shrinking and we estimate that the cupule widths of cobs in the Room 32 sample are about
25 to 35 percent smaller than they were before charring. Most cobs
from the Joyce Well site appear to have been exposed to higher temperatures or to longer heating than those from the Hidalgo County sites excavated in 1962.

Most of the corn from the Joyce Well site belongs to the Pima-Pa-
pago race and its flint variant, Onaveno. There is far less Puebloid corn
and many more cobs with small diameters. Some of these cobs with
small diameters are Chapalote or one of its closely related small-cob
forms, kinds of corn that, while very ancient in Mexico and the Southwest, continued to be grown in small amounts until recent years.

The approximate limits of the number of rows of grains and the
width of the cupule of the major races of corn from this site were measured. The cupule width limits become higher as row number decreases.
These limits roughly correspond to lines that would be determined by
measurements of cobs of various row numbers but with the same diameter of the cob itself, exclusive of the glumes. Because it is difficult to
obtain an accurate direct measure of cob diameter from entire cobs or
from fragments, we use the measure of the cavity associated with a pair
of spikelets and their grains. This measure was first used by Nickerson
(1953) and has since been used in many reports on archaeological
maize. It provides a simple method for making useful identifications and
comparisons. It is a far more sensitive indicator of cob size because it is
a measure on the perimeter of the true cob.

With our present techniques we usually have no way to find out from
carbonized material exactly what kind of corn was being grown. The
thickness of the kernels from the Joyce Well site, which can be measured
by the spacing of the cupules on the cobs, is similar to that on many
flour and flint ears. It is unlikely that many, if any at all, dents were
being grown. Some of the smaller cobs, especially those with more rows
of grains, were popcorn or exceptionally hard flints like Chapalote.
There is no way to detect sugar corn by a study of these cobs, but, since
some sweet corn was grown by the Papago not far away, some may
have been present. Very few cobs had long glumes as did the extremely
slender-cobbed types from Pinnacle Cave, also in Hidalgo County
(Lambert and Ambler 1965). This site, and U-Bar Cave, must be somewhat older than the Joyce Well site or had more conservative agricul-
turists maintaining older types of corn. The specimens came from several places in Pinnacle Cave so it is unlikely that they all represented
special caches of the slender-cobbed corn.

The Clanton Draw and Box Canyon sites have more large-cobbed
Pueblo corn, which may indicate that the Joyce Well site enjoyed less
contact with Mogollon and Anasazi peoples, had less vigorous strains
of corn, or, and this is least likely, less favorable growing conditions.
The scarcity at these last three sites of the slender and thin-glumed cobs

TABLE 3.1. Plant Remains from Joyce Well

ROOM NUMBER	PROVENIENCE	CATALOGUE NUMBER	NUMBER OF COBS	NUMBER OF ROW (Pct. of Total Cobs)				REMARKS
				8	10	12	14	
3	Room fill	2654	1		100			Small mass, about 5 ears, cupule width 5.5 mm.
3	Floor fill	597						About 16 common beans.
8	Room fill	651	6		50		50	Pars of 6 cobs, 3 laid somewhat parallel, others random. Cupule width about 5 mm.
8	Floor fill	659						Broken mesquite seed.
8	Floor contact	2657	6	17	66	17		
9	Room fill	777	77	18	56	21	5	Most ears laid parallel.
9	Floor fill	775	94	33	47	19	1	More than half were laid parallel. No husks. Ears very much like those of room fill.
12	Burial 1	2032						Small corms of unidentified monocotyledonous plant.
14	Burial 2	2623, 2634						About 118 cotton seeds.
15	Burial 1	2585						Fragments of small bottle gourd (*Lagenaria*), not carbonized; rind 2.5 mm thick. Fruit probably pear shapes between 7 and 10 cm in diameter.
15	Burial 1	2622						
16	Floor fill	2005	8	13	37	25	25	
16	Burial 1	2047						Corn husk, not carbonized.
18	Roof fall	994						Grass inflorescence, perhaps *Phragmites*, reed grass.
18	Roof matter	2542	13	8	61	31		24 *Lagenaria* rinds from medium-sized pear-shaped gourd, rind 4.4 mm thick. 35 reed grass (*Phragmites*) joints.
18	Room fill	993	88	6	41	43	10	Cobs slightly larger than those of Room 9.
18	Floor fill	997	8		25	63	12	14 *Lagenaria* rinds, about 4 mm thick, one piece round, perhaps used as a scraper. Common and tepary beans.
18	Floor contact	2012, 2647	7		29	71		*Lagenaria* fragment, 3 mm thick.
26	Floor fill	2253						Asbestos fibers.
31	Floor contact	2334	9	22	56	22		
32	Room fill	2337	44	43	47	16		Ears stacked like cordwood; metallic sheen suggests high burning temperature and relatively little air.
32	Floor fill	2335	48	35	40	23	2	Ears stacked. Some may have been slightly immature when harvested.
34	Burial 1	2550, 2552	1				100	Six cotton seeds.
34	Burial 2	2400						About 120 cotton seeds; 1 small acorn.
44	Room fill	2382	63	35	54	11		Cobs stacked.
44	Floor fill	2564	26	38	50	12		Most cobs were stacked.

so common in Pinnacle Cave may be a result of selective preservation. The small cobs and thin, long, and well-separated glumes of the Pinnacle Cave slender ears, forms of Chapalote, would burn readily, and those cobs that were carbonized and not burned away would be fragile and difficult to excavate and save.

Many of the masses of corn ears had been stacked neatly, like cordwood (Fig. 2.21), a practice that can be seen in some of Hopi homes. The husks had been removed, although a few pieces could be seen at the butts of some ears.

Many modern Indians keep different kinds of corn separate, but these piles may be close together, or a stack may have several kinds of corn in it, with those of one kind together but adjoining ears of another kind. In several visits to Hopi, we saw very little mixed corn. There were separate piles for different kinds, in a few cases even for slightly different kinds of the same color. Only one of the few stacks of corn we saw had more than one kind in it, and in that the white corn was at one end of the stack and the blue at the other end.

There is considerable diversity in the corn from the Joyce Well site, but the study of the collections from the many rooms that yielded corn makes it possible to determine that there were definite varieties and these were roughly like the kinds grown by inhabitants of some nearby sites and by the Papago Indians. An examination of the percentages of the cob and ear specimens from the various rooms (Table 3.1) shows that there is considerable difference in the corn from room to room mainly in the proportions of each kind present.

BOTTLE GOURD (LAGENARIA SICERARIA)

Several kinds of bottle gourds were grown, ranging from small and thin-shelled gourds to medium-sized ones with thick rinds. Larger gourds may have been grown, but the thick rind fragments that might have come from larger gourds were too fragmentary to use for estimates of diameters.

COTTON

About 238 cotton seeds were found. There was no evidence to show whether the seeds had been removed from the fiber or still had the fibers attached when they were carbonized. A small piece of a plain weave textile, which was attached to a 10-rowed cob (cat. no. 2571), had S-spun bast fiber warp and a slightly looser Z-spun cotton weft. Textiles usually have Z-spun cotton (Kent 1957:478, 489).

FOUR

Human Skeletal Material from the 1963 Excavation

Erik K. Reed

The collection of human skeletal remains from the School of American Research excavations at Joyce Well, Hidalgo County, southwesternmost New Mexico, represents 23 individuals. In addition, three very tiny infants—neonate or fetal—were noted in the field, but were in such extremely poor condition that they could not be collected at all. The material brought in to the laboratory is among the worst preserved I have ever seen, and Mr. McCluney is to be commended for his conscientious assiduity in salvaging these fragmentary skeletons to the greatest extent possible.

Comparatively little could be observed on most of the series, and very few measurements could be made. For instance, Burial 16/1 comprises only a portion of the left temporal, two bits of mandible, several teeth; five cervical vertebrae and three upper thoracics, one rib; distal portions of shafts of an ulna and of the radii, two metacarpals and two phalanges; pieces of both femora. Burial 17/1 consists of the right patella, the distal half of a humeral shaft; the outer half of the right clavicle, lacking epiphysis; small bits of other long bones; fragments of corpora of two vertebrae; additional smaller unidentifiable bits and chips; and a few teeth. Nevertheless, a few points of interest emerge from the study, which make the attempt worthwhile.

In the following tabulation (Table 4.1), the 23 burials collected and brought in for study are listed. For many of the specimens, there is nothing more to say beyond comments or implications in the "Remarks" column. The designations are by room numbers; the field specimen numbers, which run in somewhat different order, have been omitted in this report.

The age distribution, so far as it can be determined, appears to be

TABLE 4.1. Skeletal Remains from Joyce Well

BURIAL NO.	AGE	SEX	REMARKS
2/1	Adult	Probably female	No precise criteria of age observable; tentatively sexed only on small size of astragals, calcanea, and patellae, and of epiphyses of the femur.
3/1	Baby (6 mo. ± 6)	—	No cranial remains.
7A/1	Adult (mature, not elderly)	Male	Small (in fact, the astragals and the distal epiphyses of the femur disturbingly small, but sex pretty definite). The best skull of the series, but scarcely worth keeping; postcranial skeleton fragmentary and incomplete.
9/1	Adult	Probably male	Especially poor condition. Tooth wear somewhat greater than in 7A/l. No fragments of pelvis (or of arms and shoulder girdle).
10/1	Adult	?	A few scraps and one tooth, a premolar, wear of 7A/l.
11/1	Baby (6 mo. ± 6)	—	Few bones other than ribs.
12/1	Adult (mature, not elderly)	?	Astragalus and leg bones of 9/1 slightly smaller. Two bits of skull vault, a fragment of right temporal (and several teeth), unremarkable.
14/1	Infant ~3 yrs.	—	No determination of tooth eruption possible.
14/2	Adult	Probably female	Tooth wear quite severe, but little premortem loss. Fragmentary skull, plus fragments of scapula, a slender clavicle shaft, a small humeral head.
15/2	Adult (mature, not elderly)	Female	Fragments; no complete bones except astragals and patellae.
16/1	Adult (youthful)	Probably female	Quite small.
16/2	Adult (mature)	Probably male	Comparatively large.
16/3	Child ~11 yrs.	—	Permanent second molars erupted but unworn; canines not erupted.
17/1	Adult	?	Small to medium-sized.
17/2	Adult	?	A few small fragments, mostly cranial.
18/1	Juvenile	?	Under 18; if a female, under 15 (proximal epiphysis of the femur not united).
21/1	Adult (not elderly)	?	Nothing but a few teeth, with only moderately severe wear.
22/1	Baby (probably neonate)	—	Skull fragments only.
23/1	Baby	—	Most of a set of deciduous teeth.
24/1	Adult ~45 yrs.	Female	
29/1	Fetus, not over 6 months		
34/2	Probably adult	?	Only a few much worn fragments.
35/1	Baby (probably neonate)	—	A few bits, including maxillae, no upper teeth erupted.

fairly normal; rather surprisingly so, in fact, with so small a collection—nine (or 34.6 percent) under 1 year old (probably largely newborn, still-born, or premature), only two larger children; thirteen (or 50 percent) between 18 and 40; and only one aged 45 or over. Such a pattern is typical for life expectancy and life span in the prehistoric Southwest, as discussed in some detail in a forthcoming report (Reed n.d.).

The most important point revealed by the study of this collection is the lack of artificial cranial deformation on all skulls (nine) on which definite observation was possible. Occipital (cradleboard) deformation is characteristic of the Mogollon culture of southeastern Arizona and southwestern New Mexico from very early times—the Pine Lawn and Peñasco phases—onward (Reed 1949, 1963). Its absence in a fourteenth-century population is quite interesting and rather surprising. Possibly it corresponds to the decline or partial disappearance of the practice in Pueblo IV at Hawikuh; but this occurred, so far as present information shows, within historic times, i.e., the seventeenth century (Seltzer 1944). To the south in Chihuahua, the crania from Casas Grandes are generally deformed: "in the middle period, fronto-occipital is replaced by an oblique occipital (or lambdoid). Deformation in the latest period is still of the oblique occipital variety but undeformed skulls appear for the first time" (letter of April 8, 1964, from T. W. McKern of the University of Texas; quoted with his and C. C. Di Peso's kind permission).

The skull of Burial 7A/1, adult male, was nearly complete, but lacked most of the occipital bone. A description of it will be given as the assumed norm, and correspondences or deviations observable in other specimens will then be noted. Features not mentioned are normal human or general American Indian, or so usual in Southwestern material that I do not realize they differ in other groups of which I am not familiar.

The cranium is ovoid, small, and appearing rather low vaulted, with only moderate development of parietal bosses (submedium or slight by a North European standard). Frontal height and fullness are low, but moderate or only slightly submedium for Southwestern material. Degree of sutural complexity is low, except that the lambdoid suture is deeply serrate, as is usual in Southwestern crania. The area of the glabella is damaged slightly, but glabellar prominence was evidently moderate (by Southwestern Indian standards; low compared to North European material), with slightly developed brow ridges above the median (inner) halves of the orbits (Type I of Hooton 1930:84). The orbits are of rounded shape and only slightly sloping. The malars are small and moderately prominent. The canine fossae are well defined, concave, but not deeply excavated. The nasal aperture is small (breadth close to 25 mm). Alveolar prognathism is marked, is above the normal Southwestern average, but not extreme.

The mandible of adult male skull 7A/1 is fairly large and strong, of masculine type, with a moderately prominent square chin, wide but seemingly of median type; and manifests definite, though not pronounced, gonial eversion. Symphysis height is 35 mm.

The dentition of skull 7A/1 is marked by very severe tooth wear—not extreme by Southwestern Indian standards; shoveling of upper incisors is clearly detectable. No caries was observed. Definitely premortem tooth losses included only a few lower molars and the upper left second premolar.

A few measurements could be taken on cranium 7A/l, accurately or approximately (those with a question mark are uncertain, with one landmark missing or displaced). Vault length of approximately 175–180(?) mm and maximum breadth of 138 mm produce a mesocephalic index between 76.5 and 79.0; a basion-bregma height of 130(?) mm is medium relative to both length and breadth (i.e., metriocephalic and orthocephalic). These are normal, even typical or modal, Pueblo Indian dimensions. The minimum frontal diameter of 98 mm, however, is well above average.

Skeleton 24/1, a small adult female, about 45 years old, has the second best preserved skull, but is fragmentary and incomplete. Vault fragments are of normal thickness and suggest a small and rather low-vaulted skull, probably mesocephalic. The incomplete temporal bones have quite small mastoid processes and no development of supramastoid crests. The frontal bone is narrow, of moderate height and fullness, with no development of bosses. There is no glabellar prominence at all, and no brow ridges. Orbits are quite evidently small, round, and horizontal. The right maxilla shows medium-sized infraorbital foramina and slightly excavated canine fossae, concave but not deep. The nasal aperture is small, breadth estimated at 2-by-12, 24(?) mm, short, and very probably metrically platyrrhine. Alveolar prognathism is quite pronounced.

The mandible of skull 24/1 is very fragmentary. The dentition is characterized by extremely severe tooth wear; some caries—at least one large cavity and some premortem tooth loss—not extensive for age level, but at least a few lower molars; none lost forward of molars in left upper quarter, second premolar and first molar lost, with resorption, in right maxilla.

Other cranial material corresponds in general, so far as observable, to these two individuals. Low sutural complexity, slight or moderate glabellar prominence, small malars and rounded orbits, large infraorbital foramina and shallowly excavated canine fossae, small piriform aperture, unusually marked alveolar prognathism, square chin; and shovel-shaped incisor teeth, seem to be typical. The chin is bilateral rather than median in observable cases other than 7A/l. Gonial flare (eversion of the angles of the mandible) is exceptionally strong in large adult male jaw 16/2. The last item could prove to have some significance, being not customary in Southwestern material (though far from unknown; e.g., Hooton 1930, plates III-29 and VI-11). Another matter that at one time I thought might indicate regional differentiation is that of depth of the canine or suborbital fossa, which does not seem to check out (Reed n.d.).

Postcranial bones are no better preserved than the skulls. For example, there is just one sacrum, that of middle-aged female 24/1 (a regular five-segment one, the coccyx separate, not very curved for a female), other than barely recognizable chips. Humeri are of moderate (normal Southwestern) size and muscularity, without notable torsion. No instances of septal aperture (olecranon perforation) of the humerus was found in humeri of three individuals well enough preserved for observation. Ulnae and radii are normally muscular, unremarkable so far as observed. The only clavicle nearly enough complete to permit comment is of 16/2 (large mature adult, believed male): it is quite large (the length not measurable) but relatively slender.

Femora are small to medium sized, rather slender and not particularly muscular; without pilastering except in the case of middle-aged female 24/1. Tibiae are slender rather than stout, but not flat or "saber-shaped," with concave external surface and curved, more or less convex, posterior surface. Degree of retroversion of the tibial head is generally not determinable. Squatting facets are common, indistinct (female 24/1 and specimens 12/1 and 15/1 of uncertain sex), or well-defined but small (males TA/1 and 9/1).

No measurements were feasible on the long bones, aside from the diameter of the head of a few humeri (male 7A/1, 42 mm; females, 37 and 38 mm) and femora (male 7A/l, 44 mm; females, 38 and 39 mm). The bone lengths can be roughly approximated for specimens 7A/1 and 24/1, and are very close to the Pecos means for maximum humerus, femur, and tibia length for males and females respectively (Hooton 1930). Hence, reconstructed statures can be estimated of around 5 ft 5 inches for the male and 4 ft 11 inches or 5 ft for the female.

Other than dental troubles, those seeming to amount to somewhat less than usual, no pathological features were noted other than a little osteoarthritic growth on thoracic vertebrae of 16/2 (mature adult, presumed male) and considerable growth on the lumbar and thoracic vertebrae of 24/1 (female of about 45), which appears to be normal, virtually universal, in middle-aged Southwestern material.

The series, if this little collection of very fragmentary material can be dignified with such a title, represents, so far as it can be observed, a normal population of Southwest Plateau "Ashiwid" or Basketmaker-Pueblo type, with few divergences or unusual features. Points of special interest are (1) the surprising absence of artificial occipital deformation, (2) the seemingly consistent occurrence of moderately concave suborbital fossae and alveolar prognathism stronger than average, and (3) eversion of the gonial angles of the jaw in at least a few cases, exceptionally flaring in one. These variations of detail neither take these people out of the Pueblo or Southwest Plateau category (Seltzer 1944) nor align them with any particular subdivision within it. The material gives the impression of a quite homogeneous group.

FIVE

Ball Courts and Ritual Performance

James M. Skibo and William H. Walker

This chapter reports on the results of the excavation and analysis of the Joyce Well ball court, which is located 80 m north of the pueblo (Fig. 5.1). The role of ball courts in regional interaction is also discussed, and we provide some preliminary information on two other ball courts located within 11 km of Joyce Well (Culberson and Timberlake). Moreover, activities that took place within the ball courts at the Joyce Well, Culberson, and Timberlake communities are explored through a performance-based analysis. By focusing on the attributes of the courts themselves, the associated performance characteristics, and the ethnohistoric documentation of ball courts, we conclude that the courts had three primary functions. First, based on the accessibility, capacity, and location of the features, we argue that the courts performed as a community integrative mechanism. Second, the true north orientation of the features and their similarity in size and shape to other ball courts located to the south, suggest that a celestial-based rubber-ball game was performed in the courts that was related to fertility rituals. Third, and finally, we argue that the three ball court communities in the Bootheel of New Mexico were part of the Casas Grandes religious interaction sphere. This is based not only on the similarity in courts between the Animas phase sites and the courts located nearer to Casas Grandes, but also on other Joyce Well artifactual and architectural data.

JOYCE WELL BALL COURT

The Joyce Well ball court was not identified by McCluney's team in the 1960s. In fact, no Chihuahuan-style ball courts, which typically consist of two parallel rows of rocks and often a slight berm and interior

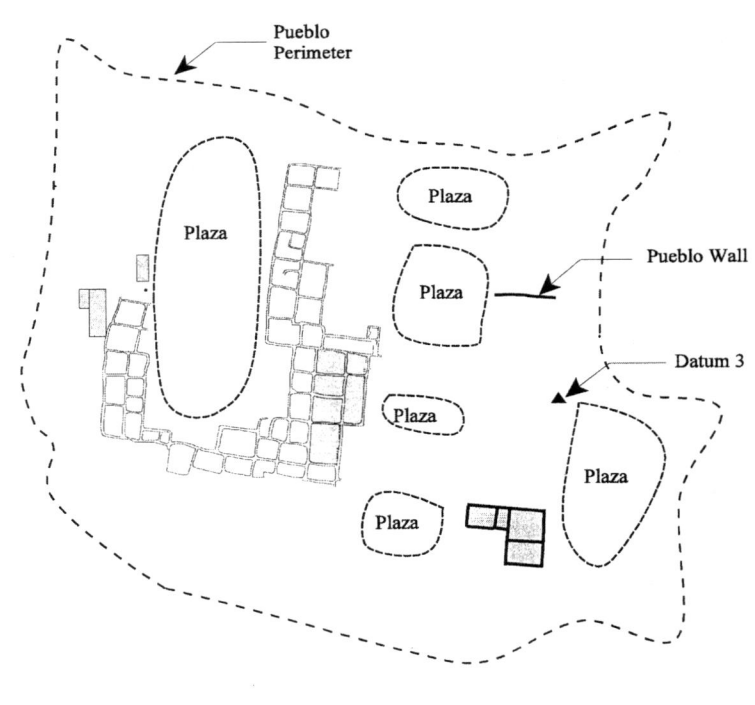

Joyce Well (LA 11823) Excavation Map

1963 McCluney Excavation

1999-2001 Excavation

0 5 10 15 20 meters

10-3-2001   *LCL*

Rock Boundary

Pueblo Perimeter

Plaza

Plaza

Plaza

Pueblo Wall

Plaza

Datum 3

Plaza

Plaza

TRUE NORTH

N 340 E440

FIGURE 5.1. The Joyce Well site showing the relationship between the ball court and the pueblo.

depression, had been recognized at that time as features within the Casas Grandes regional system. Primarily through the work of Whalen and Minnis (1996, 1999; see also Fish and Fish 1999; Leyenaar 1992; Naylor 1995; Schaafsma and Riley1999), however, many of these features have now been reported and a total of 22 have now been recorded in Chihuahua and southwest New Mexico.

The three courts in our study area are the only Chihuahuan-style ball courts located outside of Mexico. There has been some suggestion (Fish and Fish 1999:37–38) that a feature at the Ringo site (Johnson and Thompson 1963) may also be a ball court but it does not seem to be the type of court associated with the Casas Grandes interaction sphere. The feature is 26 m in diameter and it is generally oriented north-south and it appears more similar to what Whalen and Minnis (1996:736) refer to as "quadrilateral structures that *could* have been a playing field" (emphasis in original). The feature is located between two room blocks and a low adobe wall was found in the associated berm. Although there is some variability in known ball courts, the courts located in the Bootheel of New Mexico (Joyce Well, Timberlake, and Culberson) are located away from the pueblo (either directly north or south), and they are bordered by either a single or double rock wall. We cannot rule out the possibility that a ball game was played at the Ringo site but it does appear to be out of the morphological and stylistic range of the Chihuahuan-style ball courts associated with the Casas Grandes interaction sphere. Whalen and Minnis (1996) excluded these features, sometimes referred to as "corrals" (Braniff 1988), from their discussion of ball courts in the Casas Grandes interaction sphere and we will as well.

The Joyce Well court was not recognized until recently because the feature is not only quite subtle and located away from the pueblo, but it is covered by a heavy tangle of mesquite, cholla, prickly pear, and other desert scrub. Prior to excavation we visited the site on a number of occasions and typically wandered about for some time within, literally, a few meters of the rock walls made invisible by the heavy vegetation. What is more, a feature of this size is difficult to comprehend when walls cannot be easily followed and only a small segment can be seen at one time.

MORPHOLOGY

In order to understand the court's surface morphology, all vegetation was removed from the feature, which took approximately 20 people almost two days working with chain saws and hand tools. The brushing alone exposed approximately 50 percent more wall rocks and it revealed more clearly the slightly depressed court center and the associated berm. The rocks used for construction, ranging in size from 10 cm to 50 cm, were waterworn and were likely taken from the Deer Creek bed located just 100 m southwest of the feature. Ground stone fragments were also observed in the stone walls.

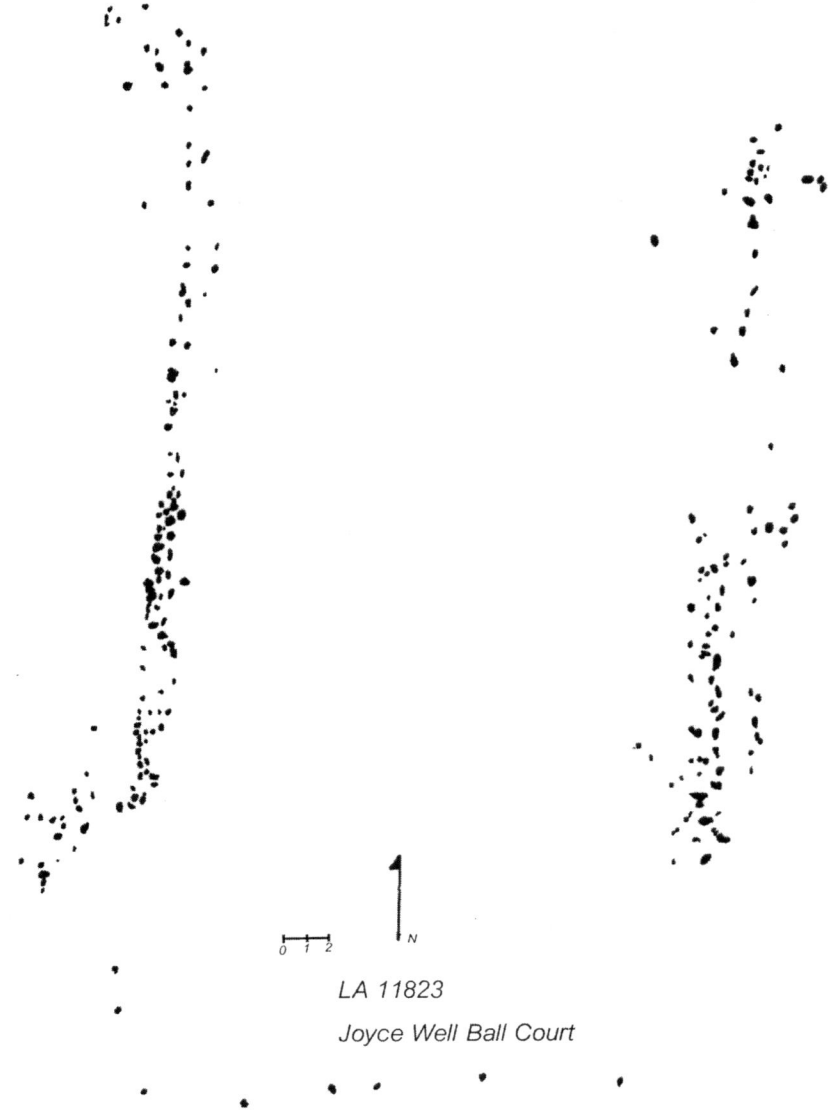

LA 11823

Joyce Well Ball Court

The Joyce Well ball court basically consists of a slight depression
bordered by two parallel rows of rocks and associated berms. A plan
view of the walls was prepared and each rock drawn to scale (Fig. 5.2).
The interior dimensions (within the two walls) are 23.5-by-35 m (822.5
sq m) and, like the pueblo, is oriented approximately true north (the
ball court is oriented 6 degrees east of north). A total of 259 rocks were
exposed during brushing (this total does not include rocks identified
during excavation). The two parallel rows of rocks with a slightly
longer west wall are clearly illustrated in Figure 5.2. We attribute this
slight difference in length to historic disturbance particularly on the
north ends of the walls. Not only have cattle grazed on the property

FIGURE 5.3. Exposed west wall of the Joyce Well ball court.

since the early twentieth century but the Joyce Well homestead was built on top of the pueblo and a historic road passes within 15 m of the court. The overall impression of the walls is that most of the rocks have been displaced from their original location most likely by cattle or pedestrian traffic.

One of the least disturbed areas occurs on the west wall just south of the midpoint where vegetation served as a sand trap to protect the walls from disturbance (Fig. 5.3). Here only the very tops of the rocks were exposed and they are set on edge in a double row. This pattern was repeated on various other less-disturbed segments leading us to believe that all of the walls had at one time the double rock row pattern. The southern end of both the east and west walls also seem to have undergone the least disturbance owing, in part, to the larger rocks used, which are less affected by cattle traffic. At the southeast corner the walls angle outward in what appears to be a crude I-shape (Figs. 5.4, 5.5). Here, it appears that the integrity of the original rock placement has been maintained owing, for the most part, to the larger rocks used.

A detailed topographic map of the ball court was also prepared by recording elevations at meter intervals. Both Figure 5.6 and Figure 5.7 illustrate clearly the central depression, east and west wall berms, and the slightly lower north and south berms. The contour intervals are only 20 cm, which exaggerates, slightly, the height of the berm and the depth of the central depression. At the east-west midline of the court the difference in elevation between the berm apex and the lowest elevation of the central depression is only about 0.30 m. Nonetheless, when standing in or near the feature the berm and central depression are unmistakable.

FIGURE 5.4. South-west "I-shaped" corner of the Joyce Well ball court.

EXCAVATION

The ball court excavation focused on defining the original surface of the court, exploring the architectural details of the walls, collecting associated artifacts, and searching for center or end court features. In all cases the excavations were quite shallow as sterile soil was encountered between 5 and 20 cm.

An east-west trench was excavated through the court midline and the original court surface was exposed (Fig. 5.8). The surface consisted of

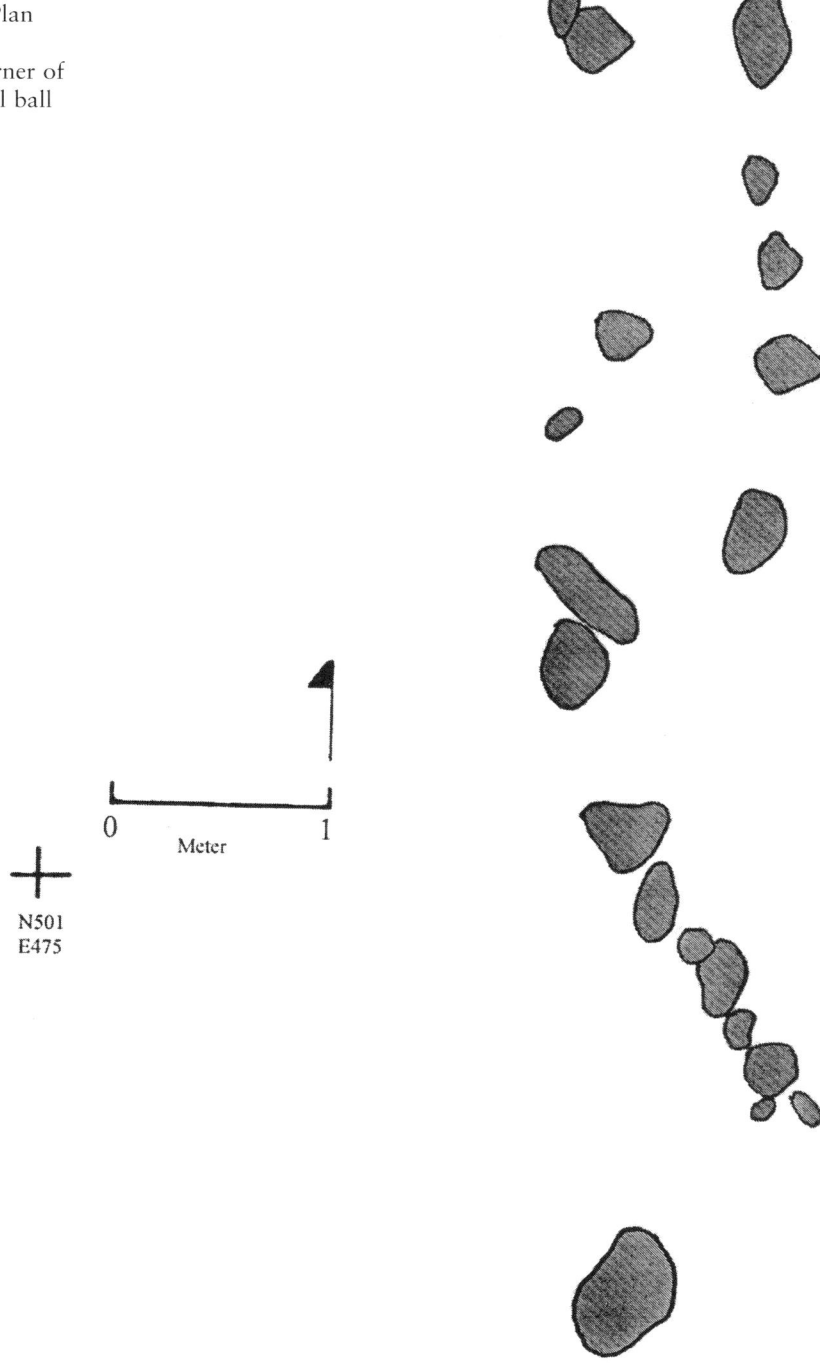

FIGURE 5.5. Plan view of the southwest corner of the Joyce Well ball court.

N501
E475

0   Meter   1

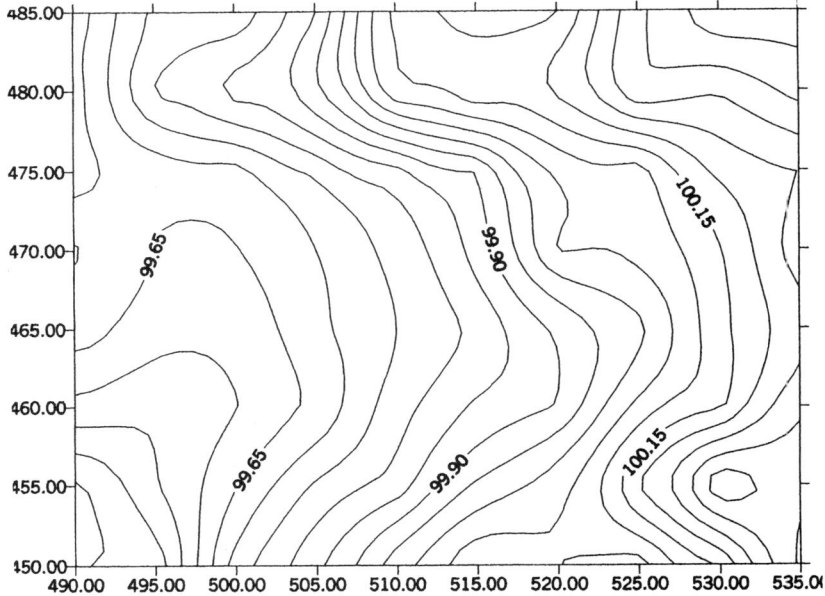

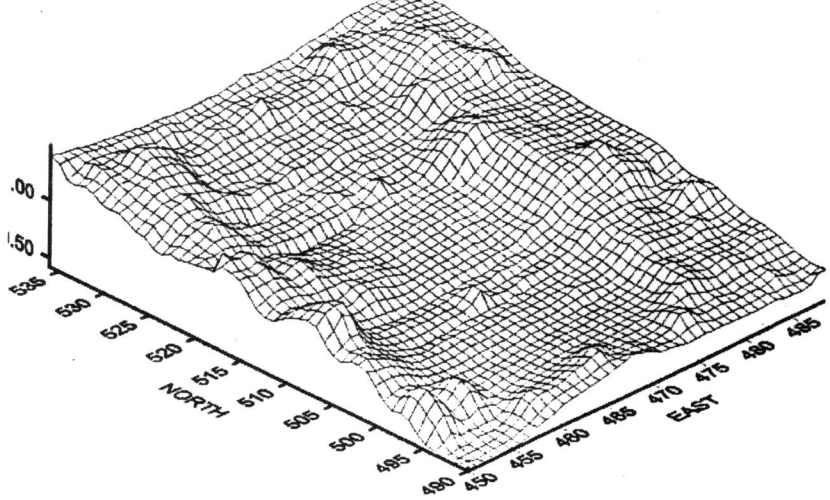

only hard-packed earth; no plastering or formal floor preparations were discovered. The trench was widened at the center of the court in the search for features but none were encountered. One important question that was not resolved to our complete satisfaction was how the ball court surface and rock wall/berm intersected because the berm had such heavy disturbance from roots and rodents. At one of the least disturbed areas (Fig. 5.9), however, it is clear that the floor rose gradually to the berm and there was not a formal wall that may have, for example, met the floor at a 90 degree angle.

Segments of the east and west walls were also exposed (Fig. 5.10) to

FIGURE 5.8. East-
west trench exposing
the surface of the
Joyce Well ball court.

determine how walls were built and to investigate whether the rocks
served as the only side-line marker or whether they were just what remained of a more formal wall made of adobe or wood. We found no
definite evidence of elaboration of the walls beyond the double row of
rocks. In some of the units, however, we did find evidence of a thin
adobe-like lens that might be the melted remnants of a low adobe
bench. Our experience in relocating the walls of the 1963 pueblo excavation suggests that walls exposed to the elements quickly melt away.

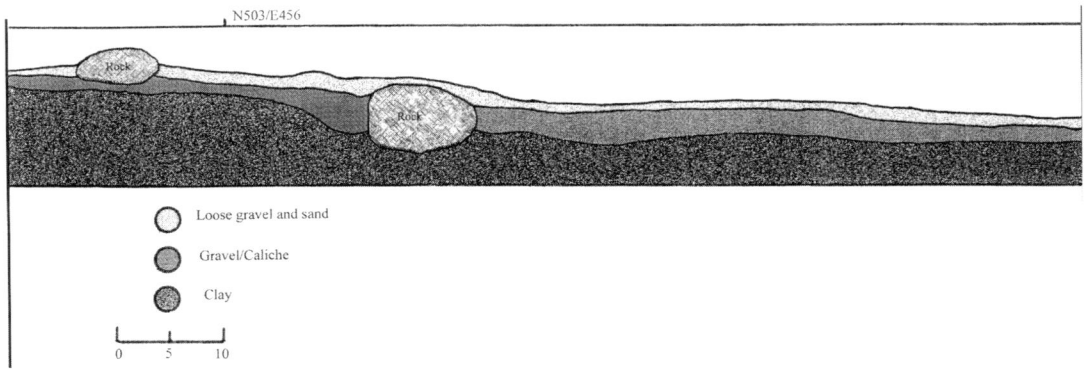

FIGURE 5.9. West wall profile of the Joyce Well ball court.

FIGURE 5.10. Exposed east wall of the Joyce Well ball court.

Our best guess at this time is that the walls simply consisted of two double rows of rocks but we cannot rule out the possibility that a low adobe bench did exist between the rocks; a low adobe wall of this type would have melted away rapidly if exposed on the berm. The only reason that the adobe walls were preserved in the ruin is because many were burned and filled before completely deteriorating.

During the excavation of the walls we did, however, find a row of buried rocks that was not associated with the previously exposed ball court rock alignment. These rocks consistently appeared about 1.5 m inside the double row rock wall. We concluded that the rocks represent an earlier court that did not have a central depression. The remodeling of the original court involved the excavation of the central depression and the associated piling of earth that created the berm and covered the rocks of the earlier court.

ARTIFACTS

Artifacts were not common in the feature but the frequency of ceramics and chipped stone was greatest in the berm. We have no evidence that the artifacts were the result of activities on the berm; they likely became concentrated in and around the walls as a result of the original excavation of the depression and, possibly, the routine cleaning of the court surface. A total of 464 sherds and 818 pieces of chipped stone were recovered from the ball court excavation. Lithic tools (limited to utilized flakes and scrapers) were dominated by utilized flakes, which accounted for 98 percent (n = 249) of the total (Table 5.1). An analysis of the debitage (Table 5.2) reveals a high percentage of broken flakes, which is an indicator of tool production (Sullivan and Rozen 1985). The lithic raw material is dominated by rhyolite (52.7 percent), which is a stone that

TABLE 5.1. Chipped Stone Tools

TOOL TYPE	FREQUENCY	PERCENTAGE
Utilized debitage	249	98.0
Thumbnail scraper	2	0.8
Side scraper	2	0.8
Flake tool	1	0.4

TABLE 5.2. Debitage Analysis

DEBITAGE CATEGORY	FREQUENCY	PERCENTAGE
Complete flake	152	20.2
Broken flake	376	50.1
Flake fragment	139	18.5
Debris	84	11.2

TABLE 5.3. Lithic Raw Materials

RAW MATERIAL	FREQUENCY	PERCENTAGE
Rhyolite	425	52.7
Chert	219	27.2
Obsidian	155	19.2
Quartz	3	0.4
Unknown	4	0.5

FIGURE 5.11. Size distribution of ball court sherds. Measurements are in millimeters.

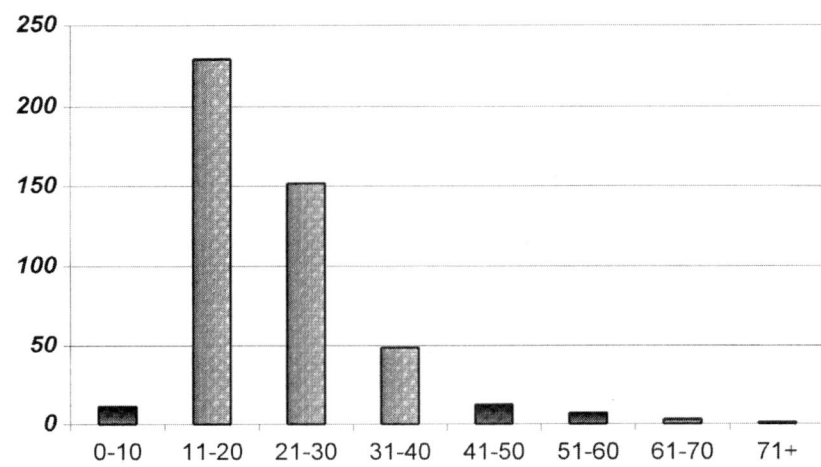

Length %

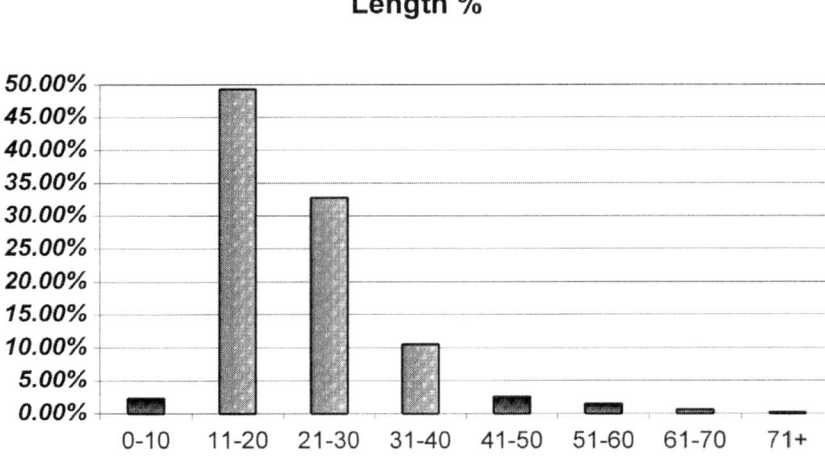

can be picked up right on the site (Table 5.3). The obsidian is all from the Antelope Wells source found several miles west of Joyce Well. At the Antelope Wells source, obsidian occurs in the form of small nodules that can be readily collected on the surface.

A total of 464 sherds was recovered. The assemblage is dominated by plain ware body sherds as only 85 pieces were painted and 12 rims were found. Very little can be said about the ceramic material because the majority of the sherds were extremely small (mean maximum length is 23.06 mm) and abraded. In fact, 47 percent (n = 216) of the sherds had extreme abrasion to the point where the surfaces were completely removed. The size distribution of the sherds also suggests that the assemblage has undergone significant trampling. Nielsen (1991) demonstrated that the size distribution of sherds after trampling will be skewed toward sizes 40 mm or less. The size distribution of the ball court sherds (Fig. 5.11) matches the Nielsen (1991) sample that was created experimentally. The vast majority of the sherds are 40 mm in maximum diameter or less, and over 50 percent of the sherds are 20 cm or less. Combined with the fact that almost 50 percent of the sherds have extreme surface abrasion, it appears that our collection of sherds have undergone extreme attrition as a result of trampling. Although cattle traffic could be responsible for some of the breakage and attrition, the majority of the sherds were below the surface and would have been protected from more recent historic activity. We argue that the majority of the sherd attrition and breakage occurred while the sherds were within the ball court. These small sherds were then swept up and deposited on the berm possibly in preparation for an event.

There is no evidence that the artifacts recovered from the excavation were produced through activities associated with the games played at the ball court. We suspect that the sherds and lithic material were deposited as a result of activity in and around the ball court when it was not being formally used. It is likely that the artifact density within the ball court is no greater than that of any area within 100 m of the pueblo. The concentration of artifacts within the berm, however, is likely the result of ball court-related activity. The berm was created by the removal of dirt from the center of the court and depositing it on the perimeter. Small artifacts of the type in our collection would simply have been deposited along with this dirt.

CULBERSON AND TIMBERLAKE BALL COURTS

Two other ball courts occur within 11 km of Joyce Well at the sites referred to as Culberson (LA 31050) and Timberlake (LA 54038) (see Fig. 5.12). Both of these courts appear at large Animas phase sites as big or bigger than Joyce Well. The Timberlake ball court occurs on Walnut Creek about 11 km northeast of Joyce Well. This court is especially interesting because, unlike the Joyce Well ball court, it is in a remarkable

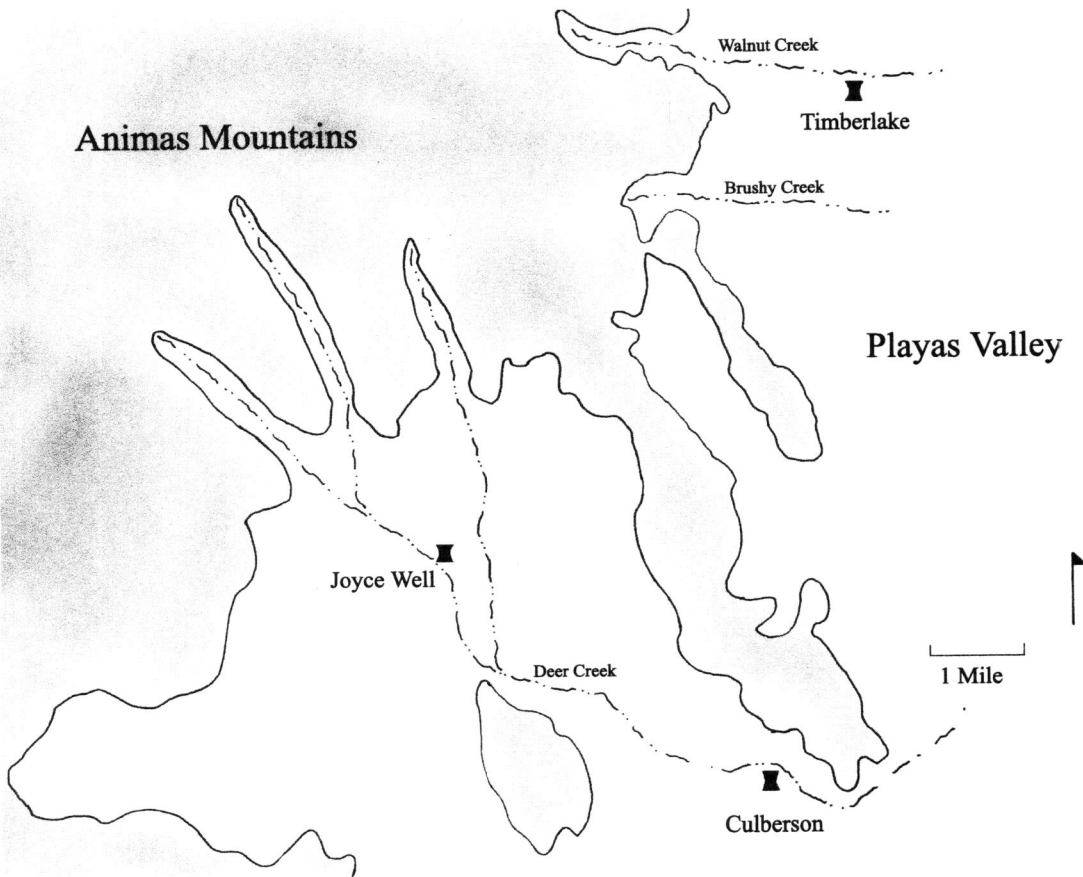

Walnut Creek

Animas Mountains

Timberlake

Brushy Creek

Playas Valley

Joyce Well

Deer Creek

1 Mile

Culberson

FIGURE 5.12. Locations of the Joyce Well, Culberson, and Timberlake ball courts.

state of preservation, which is ironic given that the Timberlake pueblo has been very heavily potted (Fig. 5.13). The court has roughly the same dimensions as Joyce Well and is oriented north (3 degrees west of true north), but beyond that there are noticeable differences in design between the two courts. Although the court consists of two parallel rock walls, like Joyce Well, the Timberlake walls consist of a single row of rocks. In most cases the rocks are still standing up on edge as originally placed (Fig. 5.13). This court also does not have a central depression, associated berms, nor I-shaped end features. Instead the court has virtually no topographic relief yet there are circular rock features located at the end of the court.

The Culberson ball court (Fig. 5.14) is also located on Deer Creek roughly 7 km southeast of Joyce Well. This feature, however, has been heavily disturbed and it is impossible to determine from the surface whether the walls were made of one or two rows of rocks. This is the smallest of the ball courts (22-by-22.5 m) but like the Joyce Well and Timberlake courts it is oriented north (4 degrees east of true north). Like Timberlake, the Culberson ball court has no associated berms and

FIGURE 5.13. Plan
view of the Timber-
lake ball court.

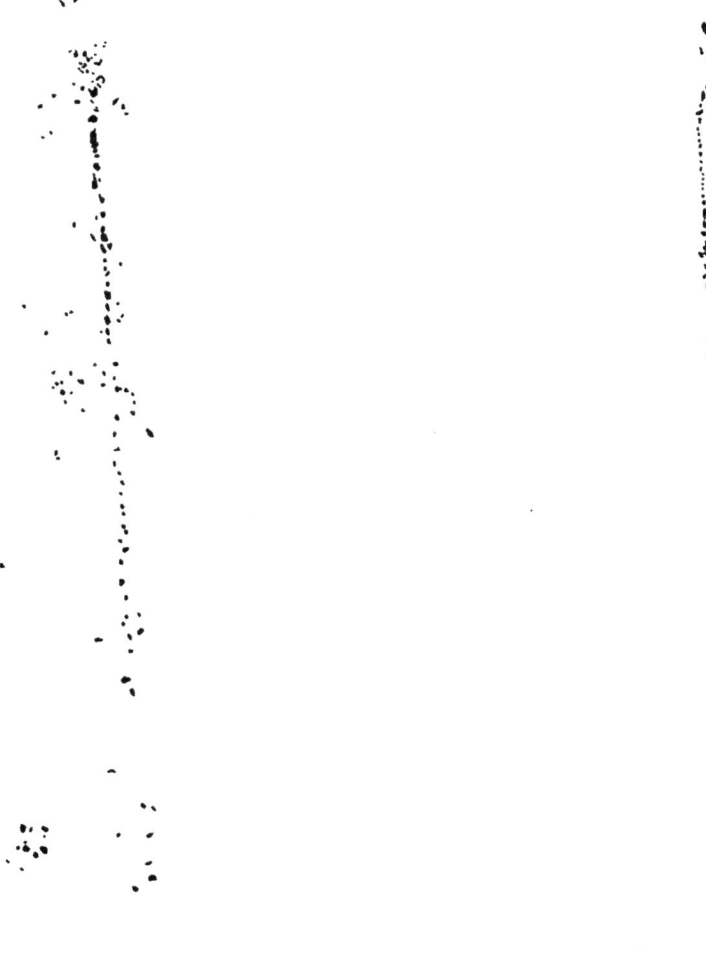

Timberlake Ball Court
LA 54038

FIGURE 5.14. Plan
view of the Culber-
son ball court.

Culberson Ball Court
LA 31050

N

1 M

depression and the court is located immediately south of the ruin. The
ball court at Joyce Well is located immediately north of the ruin. In all
three cases, however, the pattern is identical in that the pueblo is located
between the court and the stream on a north-south line.

THE BOOTHEEL COURTS

All three of the courts located in our study area are of the "simple open
variety" as discussed by Whalen and Minnis (1996). They are oriented
north-south and built simply by clearing a playing field and lining the

borders with one or two rows of rocks. In the case of Joyce Well, a central depression was excavated and piled along the borders. We think that it is significant that these Animas phase sites have ball courts but we should remember that they could be constructed and maintained with little investment of time or labor. Ten people working half a day could probably construct these courts once the area had been cleared of vegetation.

Whalen and Minnis (1996) identified a total of 21 ball courts in the Casas Grandes interaction sphere, and the location of the Culberson ball court brings the total number of courts to 22. Although the courts in the sample are similar in many ways, Whalen and Minnis (1996) also note that there is a tremendous amount of variability. Our study supports this observation. Joyce Well, Culberson, and Timberlake ball courts are located within easy walking distance, they occur at large and apparently contemporaneous pueblos yet there is a significant amount of design variability. Joyce Well has a central depression and berm with double rock walls and a crude I-shape on the southern end. This I-shape is not to be confused with the more formal type I-shaped ball courts in Chihuahua, but this is the first time that such features have been identified on the ends of the open court varieties. Timberlake does not have a depression but it does have court-end rock features and only a single-rock row pattern. Culberson is even a smaller and more simple court. Thus, even within closely associated people there is a significant amount of court variation.

RITUAL PERFORMANCE

We can never know for certain exactly what went on in these courts, but based upon the attributes of the features, their context, and ethnographic and archaeological descriptions of court function in Mesoamerica we can make some conservative inferences about court use behavior. This is accomplished by conducting a performance-based analysis of the feature, which is a strategy employed successfully elsewhere with various technologies (Schiffer and Skibo 1987, 1997; Skibo et al. 1989; Skibo and Schiffer 2001). Performance characteristics are interaction-specific capabilities that any artifact possesses. The ball court's design consists of a set of technical choices, like the size, shape, orientation, and relationship to the pueblo, that are selected based upon the feature's performance in activities during its life history. Below we outline the performance characteristics and associated technical choices important to Animas phase ball courts in manufacture and use.

Manufacturing Performance

The excavation and analysis of the ball court at Joyce Well clearly illustrates that two performance characteristics were heavily weighted by the constructors of the ball court: ease of manufacture and ease of maintenance.

1. Ease of Manufacture. Because Joyce Well has a slight central depression it would have taken longer to manufacture than Culberson or Timberlake. Nonetheless, the Joyce Well ball court would have been quite easy to manufacture. A small group of people could excavate the central depression in a short time and the simple rock alignments could be accomplished rapidly once the orientation had been determined. The floor of the ball court is simply hard packed sediment, no plaster was applied. It would have been relatively easy to make a flat playing surface. Finally, the rocks used in wall manufacture are readily available in Deer Creek, located just 100 m to the south. The rocks are of the size to be easily carried by a single individual.

2. Ease of Maintenance. The Joyce Well ball court would also be easy to maintain. Without a formal plastered floor, the court surface could be maintained simply by sweeping it clear of debris and sediment that may have come into the court between playing sessions. Likewise, the walls of the court would require little to no investment of labor. By simply righting a rock or some other equally simple task the walls could be easily maintained.

Contrast this with ball court 2 at Casas Grandes, which is deeply excavated and has formal vertical stone walls. Such a court would require a far greater investment of time and labor. Ease of manufacture would not be heavily weighted by the constructors of this court. Likewise, this court would take far more effort to maintain.

USE PERFORMANCE

The three performance characteristics important in use are accessibility, capacity, and visual performance.

1. Accessibility. The ball court is located outside of the pueblo and would be accessible to all villagers. Without a high wall or other impediments, large numbers of people could ring the court and observe the performance. The circumference of the Joyce Well court is 117 m. If you allow 1 meter per person, 117 people could ring the court standing shoulder to shoulder. Certainly 200 plus people (probably all members of the community) could be accommodated easily to observe the action if people were to sit and stand. Compare this to dances and other activities performed inside the plazas. Joyce Well has at least two plaza groups and performances conducted in each plaza would be bounded by the surrounding houses and access to the rituals could be controlled. There are, however, no natural nor created impediments to viewing the ball court activities. All people, regardless of social group or standing, could witness the performance.

High accessibility, therefore, is a highly weighted performance characteristic among the Animas phase ball courts. This is not true in some

Mesoamerican ball courts where ball courts sometimes are located in central courtyards or other areas where access is more restricted (e.g., Kelley 1991; Kowalewski et al. 1991; Santley et al. 1991). In terms of accessibility, these ball courts are more like the plazas at Joyce Well and not the ball courts.

2. Capacity. Animas phase ball courts are almost as wide as long. This is a design that would permit many people to participate in the court activity, which contrasts with some courts from Mesoamerica that may be only 3 m wide (Kowalewski et al. 1991:28–29). Ethnohistoric and iconographic information suggests that the later courts were used in games with one or two players per side. These courts are more like a long racquet ball court, whereas the Animas Phase courts are almost as big as half of an American football field. Certainly, one or two people could play in these courts, but the size of the Joyce Well, Timberlake, and Culberson courts would accommodate large numbers of people. This is a court that is designed for a game to be watched and played by many.

3. Visual Performance. The constructors of the court had many choices in design related to visual performance, which are a suite of performance characteristics that speak directly to the types of rituals performed. What we are talking about here is visual concordance. In order for this feature to play a role in a ritual performance for the Animas people it must possess a set of attributes (technical choices). Regrettably, we don't have eye-witness accounts of the ball court performances or paintings that represent some aspect of the game as they do in Mesoamerica, but we do have these technical choices as tangible traces of the rituals performed. These technical choices and performance characteristics of the feature itself are important components of the bundle of traits that make up ritual performance.

Given the great design variability between courts, the most striking visual performance characteristic of the Animas phase courts is orientation. The long axis of all three courts is oriented within just 6 degrees of true north. Of the 15 Chihuahuan courts recorded by Whalen and Minnis (1996:738), 13 were approximately oriented true north. What is more, two of the three ball courts at Casas Grandes are oriented true north (Court 3 at Casas Grandes is the exception but this court is also unique in that it is built right into the room block [Di Peso 1974; Di Peso et al. 1974:618–620]) as is the Joyce Well room block walls. This concern with cardinal direction and north in particular is clearly manifest at Casas Grandes as the walls of the pueblo are oriented in the cardinal directions, and the Mound of the Cross has four axes that point north, south, east, and west, the so-called "cardinal direction datum" (Di Peso 1974:409). Lekson (1999) has taken the notion of ritual obsession with cardinal directions to the extreme and suggests that the important sites of Chaco Canyon, Aztec, and Casas Grandes were occupied sequentially along the "Chaco Meridian," which follows a latitudinal line. Now is not the time to go into the merits of the meridian argument, but Lekson (1999) does clearly demonstrate that many people of the late prehistoric Southwest did orient their architecture in cardinal directions and that north had special significance.

This concern for cardinality is also found with ball courts outside of the Casas Grandes region. The Hohokam built 207 ball courts between A.D. 700 and A.D. 1250 and there is a tendency for the features to be oriented either north-south or east-west (Wilcox 1991; Wilcox and Sternberg 1983). This pattern of ball court orientation extends into the Mesoamerican region as well. For example, Kowalewski et al. (1991) found that prehispanic ball courts were oriented either in a near true north-south or east-west direction.

The emphasis on north and the importance of cardinal directions also can be found in the historic period (e.g., Fewkes 1892; Ortiz 1972; Parsons 1996). The true north orientation of the Animas phase courts is part of their ritual performance that links them, generally, to ceremonies that are significant throughout much of the Southwest and into Mesoamerica as well.

PRIMARY FUNCTIONS OF ANIMAS PHASE COURTS

Based upon these performance characteristics and other contextual data we argue that there are three primary functions of the Joyce Well, Timberlake, and Culberson ball courts: community integration, celestial-based fertility rituals, and integration in the Casas Grandes religious interaction sphere.

COMMUNITY INTEGRATION

We think that it is significant that the Animas phase ball courts are located at the edge of the pueblo and not built within it. Both within community and outside community ball courts are found in Mesoamerica. In each case, Gillespie (1991) argues that the ball courts function to maintain boundaries in the society. In the cases where the courts are built outside on the edge of the community, she suggests the primary function of the ball court was to symbolize the "segmentation of, and maintenance of the 'correct' distance between, sociopolitical categories" (Gillespie 1991:343). This may have been an especially important function for the Animas phase courts because sites like Joyce Well, Culberson, and Timberlake represent the largest communities in the region before or since. This is a period of aggregation in the region that brought together people that had previously been living in their own separate communities. It is likely that each plaza group at Joyce Well is composed of interrelated individuals who moved to the village as a group. In this context, the high accessibility, high capacity ball court built outside of the pueblo would have provided a means for an entire village to participate. Ball court games would then serve as a means to integrate the social groups in the village.

CELESTIAL-BASED FERTILITY RITUALS

After a review of all Mesoamerican ball games and, in fact, many pan-American games, Gillespie (1991) argues that there is a unifying, underlying function. Based primarily on iconographic representations and post-contact writings, she argues that there are two related themes to ball court games (Gillespie 1991:318–321). The first is the symbolic reenactment of the "struggle between day and night, between light and darkness" (Gillespie 1991:319). The games, therefore, symbolize the cyclical journey of the sun, moon, and other "celestial bodies." Although the details for these descriptions and relationships come from far to the south (e.g., *Codex Colombino* and *Popol Vuh*), we argue that the Animas phase ball courts shared this basic symbolic theme because of its north orientation, which is consistently associated with celestial movements. Wilcox and Sternberg (1983; see also Wilcox 1991) make a similar claim for Hohokam ball courts. They argue that "Court orientation may then have been keyed to an annual progression of calendrical ceremonies designed to keep the universe moving smoothly through its annual cycle" (Wilcox and Sternberg 1983:212).

The second but related theme of the ball courts suggested by Gillespie (1991) is agricultural fertility (see also Wilkerson 1991). A number of scholars have suggested that the game was played to insure the continuation of the cycles of the moon and sun, which are essential to agricultural fertility (e.g., Parsons 1969; Pasztory 1972, 1976, 1978). An important component of these games was the real or symbolic sacrifice of a player that symbolized the death and rebirth of the sun and moon.

INTEGRATION IN THE CASAS GRANDES INTERACTION SPHERE

Similarities between Joyce Well and Casas Grandes ball court design and other material culture are unambiguous. The people of Joyce Well had choices regarding the type of pottery, hearth design, and architectural style but they chose to mimic Casas Grandes. The rudimentary I-shape of the Joyce Well court, in particular, links it to Casas Grandes and places it within the interaction sphere. These similarities are striking and we argue that they are part of the religious interaction sphere centered at Casas Grandes.

Whalen and Minnis (1996, 1999) argue convincingly that only the communities in the central zone (within a day's walk) around Casas Grandes participated in the exchange of prestige goods. Our three sites were not under the direct political and economic control as described by Di Peso (1974) for the entire Casas Grandes system. Although our research into this area is just beginning, we argue that the pattern of behavior is most likely understood in terms of religious interaction and pilgrimage (see also Fish and Fish 1999).

Religions are often described as ideologies or systems of belief, but their organization also entails concrete interactions between people and artifacts that have practical goals and result in tangible traces in the archaeological record. Adobe pueblos enclosing a plaza, locally made Ramos Polychrome and other Chihuahuan polychromes, and ball courts represent some of these tangible clues that the people of the Bootheel were participating in this regional system. Ethnographically, in middle-range societies, local and regional cults often organize religious activities including household, community, and pilgrimage activities. Crown (1994) has argued that the spread of Salado Polychrome is best explained by appearance of a regional cult near the end of the thirteenth century. We would argue that a similar phenomenon is happening during the same time period in the Bootheel of New Mexico and the rest of the Casas Grandes interaction sphere. One difference between the Southwest Regional Cult (and for that matter the Kachina Cult) and what we call the Casas Grandes Ritual Interaction Sphere, is that the latter has a definite central place, a pilgrimage center (refer to Chapter 6 for a more complete discussion of the ritual interaction sphere).

CONCLUSIONS

This chapter reported on the first excavation of an Animas phase ball court and the discovery of a third ball court (Culberson) in the Bootheel. These are simple features made by placing two parallel rows of rocks and sometimes removing soil from the interior and piling it along the walls. Nonetheless, our performance-based analysis of the feature suggests that the ball courts played an important role in community integration and in placing the sites within the Casas Grandes ritual interaction sphere. The high capacity and high accessibility, and its location away from the pueblo suggest that the game played an important role in community integration and in the maintenance of social boundaries. The truth north orientation of the courts links it to the celestial-based fertility rituals and the site of Paquimé located 90 km to the south.

We are just beginning to understand the nature of the Casas Grandes Ritual Interaction Sphere and the role of the Animas phase sites. Our research in the coming years will focus on understanding community and household religious organization at Joyce Well and other Bootheel sites, exploring the evidence for pilgrimage behavior, and finally understanding the relationship between these Animas phase sites and Casas Grandes.

SIX

Archaeomagnetic Dating at the Joyce Well Site

Curtis F. Schaafsma, J. Royce Cox,

and Daniel Wolfman

By 1987, dating the "Casas Grandes Phenomenon" (Schaafsma 1979; LeBlanc 1980b) was becoming a major concern. More specifically, interest was mounting in the portion of this regional culture that extended into the southwestern corner of New Mexico (Schaafsma 1987). The 1984 and 1986 excavations in U-Bar Cave, Hidalgo County, New Mexico, had revived interest in the area (Phillips and Schaafsma 1987:40). Efforts were underway at Casas Grandes or Paquimé (Ravesloot et al. 1986) to revise Di Peso's dates of 1060 to 1340 (Di Peso 1974). This concern led to the April 1988 trip by Schaafsma and Wolfman to the Joyce Well site (LA 11823) in an effort to obtain archaeomagnetic samples from the hearths of rooms that Eugene McCluney reported, based on architectural evidence, to have been the last occupied at this site (McCluney 1965b). Our research goal was to obtain terminal dates for this site. Assisting on the trip were David Kirkpatrick, David Siegel, and Stephanie Daw (Fig. 6.1). Permission to excavate at the site was given by Mr. Alfredo Brenner, president of the Gray Land and Cattle Co., through his manager Mr. Fernando Rojas. Field work was done on April 5 and 6, 1988. The archaeomagnetic samples were partially processed by Wolfman and subsequently were further processed and then reexamined by Cox.

REGIONAL OVERVIEW AND MODELS: BRAND AND DI PESO

In 1935 Donald Brand made the first clear presentation about the areal extent of the regional culture related to Casas Grandes (Brand 1935,

FIGURE 6.1. Archaeo-
magnetic sample
collection field crew
(1988) at the Joyce
Well site. Left to
right: David
Kirkpatrick,
Stephanie Daw,
Daniel Wolfman, and
David Siegel.

fig. 1), which he termed the "Chihuahua complex" or the "Chihuahua
Culture Area" (Brand 1943). His original map (Brand 1935, fig. 1), site
descriptions, and regional discussions (Brand 1943) remain invaluable
resources for understanding this regional culture (Schaafsma and Riley
1999). One of Brand's students, Robert H. Lister, prepared a comprehensive overview of the culture for his master's thesis at the University
of New Mexico (Lister 1938). Di Peso essentially reiterated Brand's
"Chihuahua complex" (Di Peso 1974:1:5–9) calling it the "Casas
Grandes Archaeological Zone" (Di Peso 1974:1, fig. 5-1). Di Peso provided a detailed map presentation of the known sites in the region (Di
Peso et al. 1974:5, fig. 284-5), which relied on reconnaissance data obtained under his direction and the previous research: "The reconnaissance, when added to the data obtained by Bandelier, Sayles, and Lister
gave a rather impressive list of sites in northwestern Chihuahua" (Di
Peso 1974:1:38). Di Peso also presented an interpretive map that depicted this region as the "Casas Grandes Sovereignty" (Di Peso 1974:2,
fig. 20-2).

Both Brand and Di Peso incorporated the southwestern corner of
New Mexico, or the Bootheel, in this primary core region of the Casas
Grandes culture. DeAtley depicted the "Casas Grandes culture" in a
map (DeAtley 1980:4; DeAtley and Findlow 1982, fig. 1) that conforms
closely to Brand's and Di Peso's maps, except that it eliminates important areas on the south. She, too, incorporated the New Mexico
Bootheel in this culture, which is where she primarily worked. DeAtley
treated the sites in southwestern New Mexico as the northern frontier
of the Casas Grandes culture (DeAtley 1980). The sites in the Bootheel
have often been discussed as the "Animas phase" (McCluney 1965a).

By 1987 a number of models were beginning to appear that attempted to define what kind of culture was operating in the Casas Grandes region. Di Peso had clearly presented his mercantile model, and there was very little question about what he meant. Di Peso's interpretation of the regional culture was closely linked to his mercantile model of the origin of the culture: "It is believed that sometime around the year A.D. 1060 a group of sophisticated Mesoamerican merchants came into the valley of the Casas Grandes and inspired the indigenous Chichimecans to build the city of Paquimé" (Di Peso 1974:2:290). In a later publication Di Peso expanded upon this idea: "In the course of time, economic exploitation of the northern region from the south would bring these people together. For the Casas Grandes Valley, this came about in a dramatic way in the eleventh century when a southern merchant family, living somewhere along the Pacific coast of Middle America, inaugurated a well-designed plan of cultural conquest of the Casas Grandes sector of the Chichimecan frontier" (Di Peso 1979:11). Speaking of the Casas Grandes river drainage, Di Peso said "Along its banks, literally hundreds of satellite farming villages were built and associated with a large capital city. Together, this population formed an inland entrepôt or Mesoamerican gleaning center that thrived between A.D. 1050 and 1350. In a sense, it functioned somewhat as an elaborate, prehistoric Hudson's Bay trading post" (Di Peso 1979:12).

In Di Peso's model the "Casas Grandes Sovereignty" (Di Peso 1974:2:328, fig. 20-2) encompassed essentially the region mapped by Brand (1935, fig. 1) as the "Chihuahuan Culture." In Di Peso's model, the hundreds of villages in the region served mainly to supply the needs of the "capital city": "Foodstuffs, for example, were most likely supplied by the many satellite villages that flourished around the business center" (Di Peso 1974:2:333). There is no question that Di Peso regarded this as a hierarchical society:

> A bureaucracy arose to control the new farmers and attend to the socio-economic needs of the growing community. It consisted of a group of reigning merchants and priests who lived in exclusive quarters within the big city. . . The new lords and masters supplied the people with raw goods and marketed their finished products. (Di Peso 1974:2:18)

At the peak of this culture, in Di Peso's interpretation, the "Casas Grandes Sovereignty" governed the entire region. According to Di Peso, "In the Paquimé Phase, hundreds of mountain and valley satellite villages located within the Casas Grandes province bowed to the needs of the capital city" (Di Peso 1974:2:314–315). According to Di Peso's model, Casas Grandes itself would have to have been settled first and the surrounding "satellite" communities would have been settled later. Di Peso provided in chart form many of the dates he relied upon, the time frame he was presenting, a list of the ceramics found in the different time periods, and a brief synopsis of significant events (Di Peso et al. 1974:4, fig. 327-4).

OTHER EARLY MODELS:
MCCLUNEY, LEBLANC, AND SCHAAFSMA

Some of the earlier models maintained, advanced, or implied by others seemed to accept Di Peso's model. McCluney, for example, said "It is now known that a direct relationship existed between the settlements of the Animas and Hachita Valleys and the great Casas Grandes site in Chihuahua, Mexico" (McCluney 1965b, preface. See Chapter 2, this volume for McCluney's recent perspective) and "An inference that the Animas people were, in all respects, Casas Grandes migrants, is based upon the presence of ceramic inventory from Mexico, as appearing at the Joyce Well site" (1965b:66). McCluney's concluding remarks are in harmony with Di Peso's interpretation that the satellite villages supplied food to the central city: "Third, it appears from analysis of the total artifact assemblage of the Joyce Well Site that the inhabitants were Casas Grandes peoples who built and occupied the site for the purpose of instituting trade and to serve as an agricultural center for the raising of food crops for the Casas Grandes site located south in northern Mexico" (McCluney 1965b:86). The El Paso area stood in the same relationship according to McCluney: "There is a strong possibility that the El Paso people, like the occupants of the Joyce Well Site, were supplying corn and other food products which in turn were transported to the Casas Grandes Site" (McCluney 1965b:86). This model would assume basic contemporaneity between the Animas phase site, particularly Joyce Well, the El Paso phase, and Casas Grandes, and implicitly that Casas Grandes was settled first and that the "satellite" communities developed later. Therefore, determining the initial dates for the "satellite" communities should provide minimum beginning dates for Casas Grandes. Interestingly, McCluney maintained that Joyce Well persisted after the abandonment of Casas Grandes: ". . . after the demise of the Casas Grandes Site, the occupants of the Joyce Well site maintained their existence for a long period" (McCluney 1965b:86).

In 1977, LeBlanc defined the post-Mimbres Black Mountain phase (originally termed the Animas phase) in the Mimbres region and said "It is clear that the Black Mountain sites are closely related to the major site of Casas Grandes" (LeBlanc 1977:13). While there were similarities with Casas Grandes, there were also differences leading LeBlanc to, ". . . hypothesize that the Black Mountain phase sites are part of the Casas Grandes cultural system but were on the periphery, and the people who occupied these sites participated only to a limited extent in some aspects of the system" (LeBlanc 1977:16). LeBlanc offered dates of "between 1175 and 1300 A.D." (LeBlanc 1977:11) for these sites.

In 1979, Schaafsma maintained that a wide variety of archaeological remains spread widely throughout the Chihuahuan Desert (Schaafsma 1979, map 1) appear to have once been an interacting culture "the nature of which we can presently only suggest" (Schaafsma 1979:385).

This widespread culture included the Animas phase, Black Mountain phase, El Paso phase, La Junta focus, and the Chihuahua culture. In 1979, Casas Grandes appeared to have been established by 1060, whereas dates for the other manifestations all seemed to be after 1150. Therefore, it was logical to assume that the culture appeared first at or around Casas Grandes and "moved into the more peripheral regions after 1150" (Schaafsma 1979:387). It was already apparent that many of the peripheral regions appeared to have persisted until about 1400 whereas Casas Grandes was supposed to have been abandoned in 1340, some sixty years earlier (Schaafsma 1979:387). It was obvious that this huge region of cultural commonality could not possibly have been under the political control of Casas Grandes, especially given the indications that many sites in the peripheral regions survived the collapse of the presumed central place. By 1979, problems with Di Peso's model of a "Casas Grandes Sovereignty" were already apparent.

In 1980, LeBlanc prepared an intensive overview of the post-Mimbres periods in southwestern New Mexico (LeBlanc 1980a). He focused on the Animas and Black Mountain phases and continued to maintain that they were involved with Casas Grandes but the nature of the relationship was poorly known: "While it is generally agreed that Animas and Black Mountain phase sites are closely related to Casas Grandes, the nature of these relations is not at all clear" (LeBlanc 1980a:294). He also incorporated the El Paso phase into an extended regional culture related to Casas Grandes: "Thus, it may be more reasonable to speak of the Casas Grandes culture in the Animas, Black Mountain, or El Paso regions" (LeBlanc 1980a:296). His model for these post-Mimbres occupations indicated a spreading outward of cultural developments that originated at Casas Grandes:

> LeBlanc (1976) has modeled the sequence of events in the Mimbres area by suggesting that the depopulation of the Mimbres region by the Classic Mimbres inhabitants was a result of the rise of Casas Grandes, the latter supplying either economic or direct political pressure for such an abandonment. This model further suggests that subsequent repopulation of the Mimbres region by Black Mountain phase people was [e]ffected by people politically dominated by Casas Grandes. Elaboration of this model might suggest that repopulation of the Mimbres area was designed to exploit the available or potential resources to aid in the support of the large Casas Grandes population. Thus, the Animas and Black Mountain sites are seen as the outer fringe of the Casas Grandes system. (LeBlanc 1980a:294)

LeBlanc felt that such a close involvement of these peripheral areas with Casas Grandes would link them in time and he began to utilize his presumed upper date of 1300 for the Black Mountain phase (LeBlanc 1980a:288) to revise the terminal date of Casas Grandes itself: "Regardless of the precise nature of the relationship between the Animas and

Black Mountain phases and Casas Grandes, abandonment of both the New Mexico areas appears to be coincident with the collapse of Casas Grandes at about A.D. 1300" (LeBlanc 1980a:295). In this instance, the model was used to revise the terminal date of Casas Grandes downward at least 40 years from Di Peso's terminal date of 1340. LeBlanc was clearly only working with the data available in 1980 and specifically noted that he did not have access to data, especially dating evidence, being prepared by Suzanne DeAtley as her dissertation (LeBlanc 1980a:279).

LATER MODELS: DEATLEY, FINDLOW, AND MINNIS

DeAtley provided some of the most reliable dates available in 1980 and proposed that the sites in New Mexico dated around 1200 to about 1425 (DeAtley 1980:73). Her suggested termination date of about 1425 was well past Di Peso's 1340 termination date for Casas Grandes, and, of course, 125 years later than LeBlanc was proposing. LeBlanc allowed a slight extension of the Animas phase sites in Hidalgo County to ca. 1325, specifically at the Box Canyon and Joyce Well sites (LeBlanc 1980:288), which "is reasonable if we assume that many of the Hidalgo sites continued to be occupied after the Casas Grandes collapse" (LeBlanc 1980a:290).

DeAtley (1980) and her collaborator Frank Findlow began the trend toward the balkanization of Di Peso's "Casas Grandes Sovereignty" (Di Peso 1974:2, fig. 20-2) with their summary article of their work in Hidalgo County (DeAtley and Findlow 1980). Their model represented a definite autonomy for the various sites in what was properly called the "Northern Casas Grandes Frontier" (DeAtley and Findlow 1980:263, fig. 1). They maintained there is very little data to support the notion that the villages in Hidalgo County were in any manner simply "satellites" of Casas Grandes:

> Di Peso's (1966, 1974) work at Casas Grandes has made it clear that at one point in its development, the Chihuahua Culture was both complex and widespread, and that it may have had significant trade relationships throughout the American Southwest. . . However, while the general extent of the Chihuahua Culture, and therefore the location of the northern boundary, has been fairly well delineated through survey work, the exact relationships between the northern settlements and the Casas Grandes center are not well understood. That is, there is no information on the extent to which the northern settlements formally participated in the broadly based political and exchange activities of the Casas Grandes system. The present paper focuses on aspects of the integration of the northern frontier communities in an attempt to gain some insight into this problem. (DeAtley and Findlow 1980:263)

These authors ignore the suggestions that the Black Mountain and El Paso phases were involved in the regional culture (DeAtley and Findlow

1980, fig. 1). Their focus is on the primary region originally defined by Brand (1935), albeit truncated on the south.

The primary data they examine are detailed ceramic design analyses. Regardless of the soundness of this methodology their conclusions are clearly stated and set the trend of further balkanization:

> The region was self-sufficient, and seems to have received minimal support from the core area. Within the region [southern Hidalgo County], a weak network existed for dealing with subsistence variations by linking sites of different catchment types in different valleys . . . the integration was probably achieved through non-institutionalized means.
>
> The lack of centralization in the integration suggests that Casas Grandes had no direct role in the establishment of the frontier organization. Rather, it appears that expansion into the area occurred as an indirect result of events in the core area, and that communities developed their own general mechanisms to survive without support from Casas Grandes. (DeAtley and Findlow 1980:277)

The trend toward balkanization begun by DeAtley and Findlow was followed by a true model of rebellion presented by Paul Minnis, who took issue with the entire culture related to Casas Grandes (Minnis 1984). He made a significant distinction between "socio-political interaction and integration" (1984:186), and it was "integration" that was under scrutiny. His study is less concerned with dating than with actual archaeological data that would corroborate or deny the notion of Casas Grandes as a regional trade center and a region politically controlled by Casas Grandes. He concluded by saying: "I suggest that most of the study area (excluding northwestern Chihuahua around Casas Grandes) was not a highly integrated part of the Casas Grandes system" (Minnis 1984:190) and mentioned that "This conclusion is in substantial agreement with De Atley's and Findlow's analyses of interactions between Animas Phase populations and Casas Grandes" (Minnis 1984:190). In regard to Di Peso's mercantile model and his notion that this was the reason for the founding of the place, Minnis said "A critical examination of Casas Grandes data suggests to me that trade was not necessarily the central purpose of Casas Grandes, let alone the reason for its founding. In short, the production and distribution of most 'fancy' material goods recovered from Casas Grandes may be more a function of elite acquisition and consumption than of a mercantile economy" (Minnis 1984:191).

Considering the wide variety of models about this culture that were emerging by 1987 and the range of dates that were being proposed, Schaafsma and Wolfman decided that an effort should be made to determine an upper or terminal date for the sites in southern Hidalgo County. Partly the choice of sites in this area was made because of ease of access (permission to work at the sites was a matter of obtaining permission from the ranch owner) and partly because a review of

McCluney's manuscript on the Joyce Well site indicated that archaeo-magnetic dates from late rooms at the site could be obtained by simply removing the dirt covering the previously excavated hearths. There would thus be no need for a full excavation of undisturbed rooms to get to the hearths.

ARCHAEOMAGNETIC DATING

Archaeomagnetic dating is a very useful tool that can be used to help define and delineate the many chronological interpretations that are proposed for any given region. The basis of previous timelines that have been presented for the "Casas Grandes Phenomenon" show how important chronologic control actually is to establish the legitimacy for any given time frame. Archaeomagnetic dating is one of several chronologic tools available to the archaeologist. Although not ideal for all applications, it can rival the precision of both ceramic and radiocarbon dating in many instances, and it is usually the best method for dating abandonments. It has the added advantage of contributing to questions of contemporaneity, both within sites and across regions. The Joyce Well site is an excellent example of the application of archaeomagnetic dating to a case where other dating techniques fail or lack the desired precision, and where results are applicable to the cultural events of a region as well as simply a site.

THE JOYCE WELL SITE

McCluney had visited the Joyce Well site in 1962 (McCluney 1965b:10) and obtained a surface sherd collection that indicated this was one of the latest Animas phase sites (McCluney 1965a:46). McCluney conducted excavations at the site in 1963 and upon considering the ceramics offered dates of approximately A.D. 1250 to 1400 for the overall occupation: ". . . the site, based upon the correlation of pottery and past information from similar sites, denoted that the initial occupation of the site was made ca. 1250–1275 A.D. The abandonment, probably periodical rather than en masse, began about 1300–1350 A.D. and continued until final abandonment was completed ca. 1380–1400 A.D." (McCluney 1965b:85). McCluney obtained radiocarbon dates from a pile of stacked, burned corn in Room 24, which he did not mention in his manuscript—probably because they were so far from agreement with the ceramic evidence. On the other hand, he communicated these dates to Di Peso who published them and relied upon them for the definition of his Robles phase (Di Peso et al. 1974:4, fig. 327-4). The midpoints of these dates as Di Peso published them are A.D. 1620, A.D. 1595, and A.D.1565. DeAtley recognized that these dates had not been corrected for isotopic fractionation (DeAtley 1980:70). When corrected at that time, these dates became A.D. 1365, A.D. 1340, and A.D. 1330 (DeAtley

1980, table 3). Since they probably all represent the same burning event, it is sensible to average them giving a date of A.D. 1340 ± 68 (DeAtley 1980, table 3). Room 24, therefore, became one of the targets for obtaining an archaeomagnetic date from the hearth.

On the basis of door closure patterns, McCluney had identified a set of rooms that were the latest occupied: "By observing the filled doorways, we were able to tell whether the doorways were plastered in from the interior of the room or from an a[d]joining room. It became clear that the last rooms to be abandoned at the Joyce Well site were the centrally located rooms 11, 12, 15, 18 and 21" (McCluney 1965b:27). Given the fact that the archaeomagnetic procedure gives dates for the last time a fire was made in a feature, it seemed reasonable to locate the hearths in the above rooms and obtain archaeomagnetic dates for the last time fires were made in the latest rooms in this late site.

Four archaeomagnetic sets were recovered from the Joyce Well site by Wolfman and Schaafsma ("sets" of individual specimens or samples are measured to derive archaeomagnetic dates). Utilizing McCluney's site map, the rooms in question were readily located. Time at the site was used to obtain archaeomagnetic samples from Room 24 (source of the radiocarbon dates), and from three of the architecturally late rooms, 12, 18, and 21. The three architecturally late rooms are contiguous, adjoining each other at the corners (Room 20 separates them). It thus seems likely that they were all occupied about the same time. Room 18 was chosen because the hearth was an elaborate raised feature (McCluney 1965b:29–30). All of the sets were collected from central hearths or firepits.

Room 12

The firepit in this room was readily located and was in very good condition. It is excavated into the floor of the room and should date the latest occupation at the site.

Room 18

The raised hearth in this room was associated with the last floor made in the room, and therefore should be one of the latest features used at the site. Unfortunately, this feature had deteriorated and the hearth was not usable. Testing around the eroded hearth located another hearth underneath it and slightly askew toward the south. This hearth was associated with a lower floor about 15 cm below the upper floor and this hearth may or may not be associated with the walls of Room 18, but it is given that designation. This lower hearth was cleared enough to obtain an archaeomagnetic date. There is no ceramic or architectural evidence for determining how much older the lower floor is (and therefore, the lower hearth) than the upper floor based on our limited work. It is

FIGURE 6.2. Hearth in Room 21 prior to removal of the sample.

not simply an earlier hearth that was later remodeled, but it is distinctly earlier than the raised firepit encountered and illustrated by McCluney as part of the final occupation of the site (McCluney 1965b). A small hole sufficient to obtain the sample was opened. However, the fill of the lower hearth was collected, and it should be possible to obtain carbonized annuals from the fill to derive a radiocarbon date. This has not been done, but could be done.

Room 21

As with nearby Room 12, the firepit shown on McCluney's map in the west center of this room was readily located (Fig. 6.2). The hearth is associated with the final floor, and also should date the latest occupation at the site.

Room 24

Room 24 is separated from Room 21 by one room, Room 22. Mc-Cluney did not include this room in his set of architecturally late rooms, being a slightly earlier component at the site. This is the room in which the carbonized corn had been recovered for radiocarbon assays. Room 24 is certainly near the rest of them, and may well be essentially contemporaneous. The firepit was easily located to obtain an archaeomagnetic sample. This hearth and the set from Room 18 afforded the possibility of contrasting the abandonment dates obtained from Rooms 12 and 21 with those of possible earlier occupations at the site.

Tentative results from the measurement of these sets have been previously reported (Schaafsma and Wolfman 1989; Phillips and Carpenter

TABLE 6.1. Archaeomagnetic Results Initially Reported from Joyce Well (LA 11823)

SAMPLE NO.	ROOM	FEATURE	VGP LATITUDE	VGP LONGITUDE	α_{95}	δ_p	δ_m	N	DEMAGNETIZA-TION STATUS	DATE RANGE(S) (A.D.)
JW 351	24	Hearth	86.6	190.6	4.3	4.1	5.9	8/7	200 Oe	
JW 352	21	Hearth	83.5	212.3	2.6	2.7	3.8	8/8	NRM	1220–1295 1335–1390
JW 353	18	Subfloor hearth	83.9	195.4	1.6	1.6	2.2	8/7	NRM	1245–1285 1330–1360
JW 354	12	Hearth	84.4	223.7	1.5	1.5	2.1	8/8	NRM	1335–1390
JW 352, 354			83.6	217.6	1.4	1.4	2.0	16/15	NRM	1345–1370
JW 352, 353, 354			84.1	212.6	1.0	1.0	1.4	24/20	NRM	1345–1370

1999) and differ slightly from the results reported here. The previously reported archaeomagnetic dates (Table 6.1) represent the initial results prior to a more extensive demagnetization and remeasurement program carried out in 1990. Shortly after collection in 1988, three of the four sets (JW352, JW353, and JW354; see Table 6.1 for concordance with the room numbers) were measured at natural remnant magnetism (NRM) with no subsequent demagnetization. The fourth set (JW351) was measured at NRM, and a pilot group of four specimens was demagnetized and remeasured through 50 Oe, 100 Oe, 150 Oe, 200 Oe, and 300 Oe steps. A second demagnetization and remeasurement was done with the four remaining specimens at 200 Oe. Demagnetization removes weakly held magnetic vectors that may not be the vector of interest in dating the last use of the sampled feature. The two-stage demagnetization process applied to JW351 (Room 24) represents a demagnetization protocol that reduced measurement time but that is no longer used because it compromises the analysis of demagnetization data. The three sets that were not demagnetized in 1988 were remeasured in 1990. At this time a full demagnetization was instituted in which all specimens were demagnetized together through the above steps. The archaeomagnetic results reported in the paper by Schaafsma and Wolfman (1989) and repeated by Phillips and Carpenter (1999) are preliminary and do not represent a full measurement process. Unfortunately, the original data that was presented contained several errors. Some information given was mistakenly taken from different outlier results at the same demagnetization level for the given archaeomagnetic set. The site location (latitude and longitude) also differed slightly from the true location, as well as the site declination (obtained using MagCalc ver 2.1 [1995], D. A. Schneider). After accounting for these differences, the archaeomagnetic results and date ranges reported here are "better" (Table 6.2) and should be considered replacements for the original tentative results.

Table 6.2. LA 11823 Archaeomagnetic Set Results

Sample No.	Room	Feature	Inclination	Declination	VGP Latitude	VGP Longitude	α_{95}	δ_p	δ_m	N	Demag. Status	AM Date range	Collector
JW 351	24	Hearth	51.3	353.9	84.8	169.1	3.0	2.7	4.0	8/6	200	A.D. 1260–1345 A.D. 930–1030	DW
JW 352	21	Hearth	55.8	354.6	83.4	210.3	2.6	2.7	3.8	8/8	NRM	A.D. 1330–1400 A.D.1220–1290 A.D. 950–1050	DW
JW 353	18	Subfloor hearth	54.1	353.6	83.7	193.8	1.6	1.5	2.2	8/7	NRM	A.D. 1240–1285 A.D. 1330–1360 A.D. 980–1030	DW
JW 354	12	Hearth	53.9	356.2	85.6	205.2	2.0	1.9	2.8	8/8	600	A.D. 1325–1360 A.D. 1240–1295 A.D. 940–1020	DW

ARCHAEOMAGNETIC DATING PROCEDURE

Archaeomagnetic dating involves the comparison of a past virtual geo-magnetic pole location (VGP) from a burned feature with the reconstructed path of VGP variation through time. Specimens are carved from the feature (Fig. 6.3), the magnetic vector for each specimen is measured and remeasured following a demagnetization protocol, and the sets of specimen vectors are statistically combined to form a single VGP result. A result consists of a centerpoint estimate and an oval that represents a 95 percent confidence interval ($\propto 95$) that encompasses the individual specimen VGPs. If the specimens of a set yield coherent VGP directions, the $\propto 95$ (measured in degrees) is small and can support precise date interpretations. If the specimens of a set yield dispersed VGP directions, the $\propto 95$ is large, usually due to being weakly oxidized, resulting in a less precise date interpretation. $\propto 95$ values greater than 4.0 not only have large error terms, but such dispersed results also raise the possibility that the individual specimen vectors include magnetic components that are unrelated to the burning event of interest. The Museum of New Mexico's Archaeomagnetic Dating Laboratory (ADL) generally avoids interpreting results from sets with $\propto 95$ values greater than 4.0. Once set results are calculated, archaeomagnetic dates are interpreted from the location of the centerpoint and the size of the oval of confidence relative to the relevant VGP curve or curve segments.

VGP dating curves are assembled from measurement data rather than being derived from theoretical models of the earth's magnetic field. Because the earth's magnetic field is not symmetrical, VGP curves must be constructed region by region. The Southwest region has the best documented VGP curve, but the curve is constantly being refined as additional data are accumulated. There is no single agreed-upon construction technique for dating curves (Wolfman 1984), and the existing dating curves have definable strengths and weaknesses (Cox and Blinman 1996; Wolfman 1984, 1990). The ADL uses the Wolfman curve for date estimation in the A.D. 1000–1450 period. Experience with thirteenth- and fourteenth-century samples indicates that this curve is a more accurate reflection of VGP movement for this period than any of the other curves constructed for this time period (DuBois 1989; LaBelle and Eighmy 1995).

Several procedures may be used to interpret dates from VGP results. Standard ADL procedure consists of (1) defining relevant curve segments and (2) estimating date ranges from the overlap between the set result and each relevant calibrated curve segment. Since VGP curves loop and overlap, there may be several date interpretations for any single sample VGP. Technically relevant curve segments are those that are overlapped by or immediately adjacent to a sample's oval of confidence. The archaeologist may have independent dating information that can be used to eliminate one or more technically relevant curve segments. For

example, at least two late Archaic curve segments cross the curve segment for the mid nineteenth century, but the Archaic date possibilities would not be relevant to the dating of a Territorial period homestead. Once relevant curve segments have been identified, a date range is calculated for each segment. The oval of confidence is manually moved so that the centerpoint is superimposed on the closest point of the curve segment. The points where the oval intercepts the curve determine beginning and end points for the date range, rounded to the nearest five years outside of the intercepts. The resulting date ranges are conservative relative to other dating techniques, but we presume that the greater conservatism makes up for weaknesses in the curve calibration. The relevant curve segment for the Joyce Well site is considered to be between A.D. 1200 and 1450 and so all date interpretations will be concerned only with this time period, thus excluding all other date interpretations.

ARCHAEOMAGNETIC DATING RESULTS

The archaeomagnetic set from Room 21 (JW352) was measured at NRM and again after demagnetization steps through 300 Oe. The VGP at NRM was chosen as the best result and is plotted in Figure 6.4. The confidence oval intercepts two relevant curve segments, one with a nearest point at A.D. 1260 and the other with the nearest point at A.D. 1355. The date ranges for these segments are A.D. 1220–1290 and A.D. 1330–1400, respectively. The room in which this set was collected represents the terminal occupation of the site, and the date range of A.D. 1220–1290 is inappropriate given McCluney's (1965b) assumption that this was one of the last occupied rooms.

The archaeomagnetic set from Room 12 (JW354) was measured at

FIGURE 6.4. Plot of
Room 21 (sample no.
JW 352).

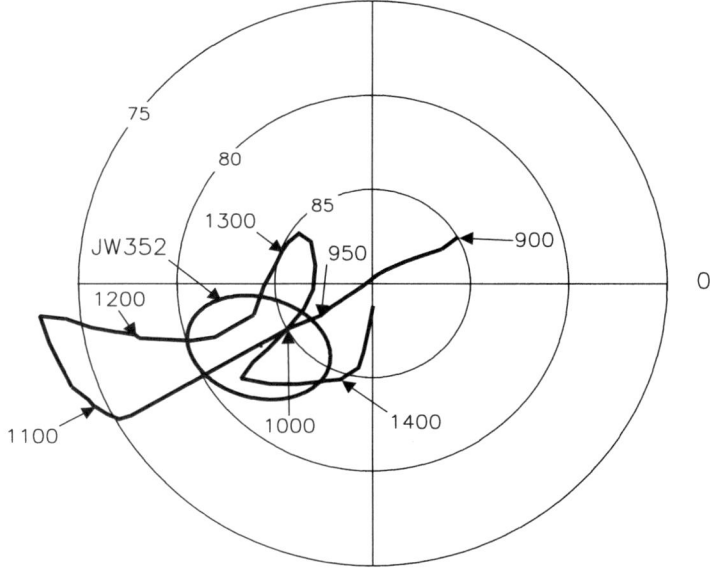

Southwest Polar Curve A.D. 900-1425+

NRM and again after 50 Oe demagnetization steps to 200 Oe and again after 100 Oe demagnetization steps to 800 Oe. The result after demagnetization at 600 Oe was determined to be best, and that result is plotted in Figure 6.5. This result overlaps two relevant segments of the VGP curve, one with a nearest point at A.D. 1270 and the other with a nearest point at A.D. 1340. The associated date ranges are A.D. 1240–1295 and A.D. 1325–1360, respectively. As with the archaeomagnetic set from Room 21, the sampled feature represents the terminal occupation of the site and the earlier date range can be eliminated from consideration based on the other dating information from the site.

The archaeomagnetic set from Room 24 (JW351) was measured and demagnetized shortly after it was taken. It was subjected to an abbreviated demagnetization protocol, with the probable loss of information as a result. Additional demagnetization may possibly improve the result, but the original data are used here. This set was measured at NRM and after demagnetization at 200 Oe. The result at 200 Oe is the better of the two, however it has an $\propto$95 of 4.3 after one outlier is thrown out (Fig. 6.6), which is slightly higher than the 4.0 that is normally considered acceptable by the ADL. The large oval intersects the A.D. 1235–1365 and A.D. 1390–1425+ curve segments. Since this archaeomagnetic set comes from a room that was abandoned prior to the final occupation of the site, the earlier range is probably the most relevant. The result after a second outlier is eliminated is a more acceptable $\propto$95 of 3.0. This second outlier is not "statistically" thrown out, but its tendency is in moving away from all the other specimens. The resulting

FIGURE 6.5. Plot of
Room 12 (sample no.
JW354).

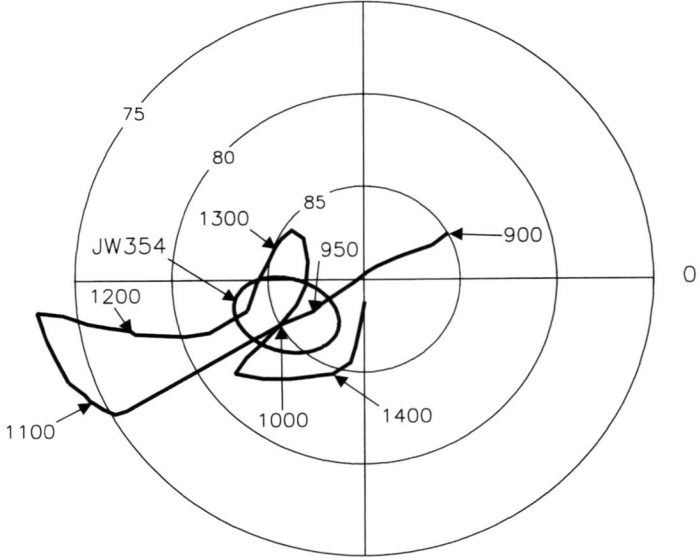

Southwest Polar Curve A.D. 900-1425+

FIGURE 6.6. Plot of
Room 24 (sample no.
JW351).

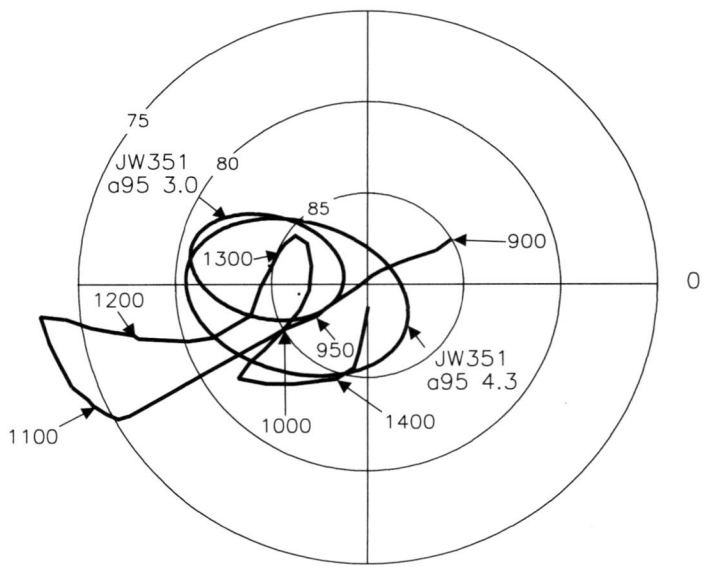

Southwest Polar Curve A.D. 900-1425+

position of JW351 (Fig. 6.6) overlaps one relevant segment of the VGP curve with a nearest point at A.D. 1290. The associated date range is A.D. 1260–1345. The date estimates from both interpretations are consistent with the results from Rooms 12 and 21 in being before the final abandonment of the Joyce Well site.

FIGURE 6.7. Plot of Room 18 (sample no. JW353).

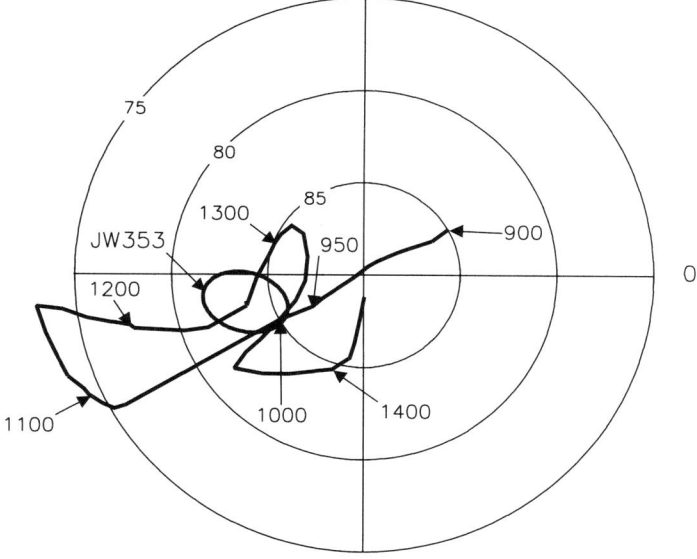

Southwest Polar Curve A.D. 900-1425+

The final archaeomagnetic set is from the hearth in Room 18 (JW353). The specimens were measured at NRM and after standard de-magnetization steps to 300 Oe. The best result was at NRM, and it is plotted on Figure 6.7. The result also overlaps two segments of the VGP curve, but it differs from the Room 12 and 21 results in that the center-point lies on the earlier segment of the curve, at A.D. 1265. The sample only barely overlaps the later segment, and the closest point on that seg-ment is at A.D. 1345. The date ranges for these segments are A.D. 1240–1285 and A.D. 1330–1360. This result applies to a hearth that was sealed over by the final construction phase at the site, and it could be at least a generation earlier than the results reported for Rooms 12 and 21. Although there is little independent evidence at present for the date of the initial occupation of Joyce Well, the early date range of this set does fall within the ceramic date range proposed by McCluney (1965b). The later range cannot be eliminated from consideration, but the earlier range is somewhat more likely.

ARCHAEOMAGNETIC DATING DISCUSSION

The archaeomagnetic dates agree reasonably well with the small amount of other chronological data available from this site, as well as the expectations of McCluney (1965b). The use of obsidian hydration dating is suspect, but none of the dates provided for this site (Carpenter 1985, Chapter 7) support or invalidate any of the dating possibilities presented. Three radiocarbon assays are available for burned corn excavated from Room 24. These were reported by Di Peso (1974) and were then reexamined and interpreted by DeAtley (1980). DeAtley's calibrated dates were then used by Phillips and Carpenter (1999) in their review of regional chronology. These dates are applicable and are revisited here for comparison with the archaeomagnetic date estimates.

Radiocarbon dating is similar to archaeomagnetic dating in that an independent calibration process is required to translate the radiocarbon "age" of a sample into a calendric date estimate. The process of calibration is being continually refined as additional data and programs are developed, and the previous interpretations of the Joyce Well dates are based on older versions of the radiocarbon calibration curve. We have applied the 1996 version of the University of Washington Quaternary Isotope Lab Radiocarbon Calibration Program (revision 3.0.3d) to the radiocarbon ages reported by DeAtley as corrected for the isotopic fractionation associated with corn. The corn is all from the same cultural event (the burning of the contents of Room 24), and the burning occurred before the final construction episode at the site based on McCluney's evaluation of architectural sequences. The intercepts for the three dates all fall within the A.D. 1310–1401 period. Error terms for the three dates overlap in the A.D. 1294–1427 period at 1 sigma and in the A.D. 1230–1479 period at two sigma. The radiocarbon calibration for this period records several oscillations, so that two of the dates have three intercepts.

Although the archaeomagnetic result from Room 24 is presented with two different interpretations, it and the other archaeomagnetic results from the site can be used to interpret the radiocarbon date results. The centerpoint of the larger Room 24 result falls on the VGP curve at A.D. 1330, and the oval of confidence encompasses the curve to about A.D. 1365. The large oval also intersects a later curve segment, with a centerpoint of circa A.D. 1420 if the result were moved to the nearest point on this curve. The archaeomagnetic results from Rooms 12 and 21 do not accommodate post-A.D. 1400 interpretations, suggesting that the only valid interpretation of this Room 24 result also must be in the early fourteenth century. The second and smaller result for Room 24 foreshortens the larger date range by approximately 20 years on each end and has no second date interpretation. The same logic applies to the post-A.D. 1375 centerpoints of the radiocarbon calibrations. These late interpretations of the radiocarbon results are improbable given the archaeomagnetic dates.

There are still too many uncertainties in these interpretations for a final conclusion about the dating of the Joyce Well occupations, but the most parsimonious interpretation can serve as a model for additional testing. The earliest occupation at the site probably falls within the 1250–1275 time period. The subfloor hearth in Room 18 may even be a feature relating to the first occupation of Joyce Well. The burning event documented in Room 24 probably occurred sometime in the first third

FIGURE 6.8. Plot
comparing Room 12
at Joyce Well with
sample from Casas
de Fuego, Chihuahua
(sample nos. CF1174
and CP1175).

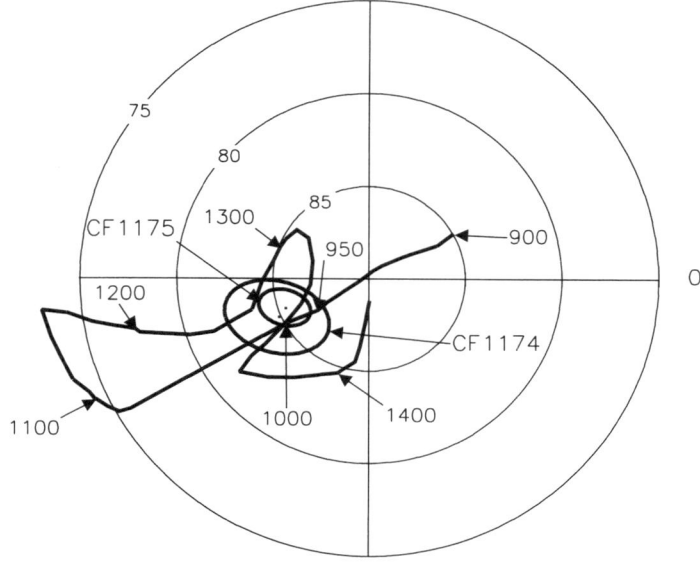

Southwest Polar Curve A.D. 900-1425+

of the fourteenth century, followed by the final room block construction
and ultimate abandonment by no later than the 1360s as seen from
Rooms 12 and 21. This model should be testable with the collection of
additional archaeomagnetic sets from Joyce Well.

A final and important caveat about this chronologic reconstruction is
that future revisions of the archaeomagnetic dating curve will result in
slight revisions of the chronology for Joyce Well, just as radiocarbon
dates are revised. The coincidence that all of the Joyce Well centerpoints
lie on segments of the existing curve suggest that the curve is robust for
this period. Although there is always a chance that an individual result
may lie outside of the error ellipse, the strong agreement among the four
results lends confidence to both the calibration curve and this interpretation.

As a final contribution, it is interesting to note the similarity between
the archaeomagnetic results from Rooms 12 and 24 and two archaeomagnetic sets recently collected from the Casas de Fuego site 19 km
southeast of Paquimé in the Arroyo Seco (presented here with the permission of Timothy Maxwell from the Office of Archaeological Studies,
Museum of New Mexico). The Casas de Fuego results are from burned
rooms that may represent abandonment of the site. These results are
plotted on Figure 6.8. The more precise result of the two sets (CF1175)
lies entirely within the oval for Room 12. The second result (CF1174)
is slightly off of the VGP curve, but when moved to the curve gives a
nearly identical date range as the abandonment dates for the Joyce Well
site. This coincidence of abandonment dates also compares with

McCluney's (1965a) estimation of the abandonment of Clanton Draw at approximately the same time and the abandonment of Box Canyon slightly earlier (these dates based solely on ceramic assemblages). The coincidence of these abandonments raises the possibility of a contemporaneous regional settlement pattern disruption. There is no evidence that the abandonment of the Hidalgo County sites was anything other than peaceful (McCluney 1965a, 1965b), and regional environmental factors such as prolonged drought may be worth investigation. At this stage, the causes of the collapse of the "Casas Grandes Interaction Sphere" are quite unknown (Schaafsma and Riley 1999).

Archaeomagnetic dating is an important chronometric tool in Southwestern archaeology. It can address many problems of chronological control, including site abandonments, and can establish a foundation for the study of regionwide cultural phenomena. There are discrete time periods where archaeomagnetic dating can also deal with problems that are less successfully addressed by radiocarbon dating due to the vagaries of the radiocarbon calibration curve. Joyce Well is an excellent example of the remedial application of archaeomagnetic dating, as burned features are often left undisturbed and backfilled by excavators, resulting in their preservation for future sampling.

ACKNOWLEDGMENTS

We are grateful to Mr. Alfred Brenner and Mr. Fernando Rojas for kindly extending us permission to work on the Gray Ranch in April 1988. We also very much appreciate the assistance of David Kirkpatrick, Stephanie Daw, and David Siegel in the field work. Eric Blinman assisted in assembling the final manuscript, and we are much indebted to him for that.

ANIMAS PHASE
RAMOS
POLYCHROME
JOYCE WELL
CARRETAS/HUERIGOS
POLYCHROME

SEVEN

The Animas Phase and Paquimé

Regional Differentiation and Integration at Joyce Well

JOHN P. CARPENTER

For many years now, the Animas phase communities of southwest New Mexico have been interpreted as a peripheral manifestation linked to developments at Paquimé, located approximately 100 km south of the international border, in Chihuahua, Mexico (DeAtley 1980; Di Peso et al. 1974; Douglas 1992, 1995; Gladwin and Gladwin 1934; Kidder et al. 1949; Minnis 1984, 1989; Ravesloot 1979; Sayles 1936a). Paquimé was, unquestionably, an important regional center with single and multistoried puddled-adobe room blocks, ceremonial architecture including ball courts and platform mounds, a complex water storage and distribution system with reservoirs, subterranean conduits, and a walk-in well (Di Peso et al. 1974).

According to Di Peso (1974), the meteoric growth of Paquimé as a regional center could be attributed to the intrusion of Mesoamerican merchants, or *pochteca,* specializing in long-distance exchange, who transformed a small Chihuahua Mogollon hamlet into an international port-of-trade. Pailes and Reff (1985), Plog et al. (1982), and Whitecotton and Pailes (1986) follow Wallerstein (1974) in elaborating upon Di Peso's concept of Mesoamerican interaction, suggesting that Paquimé was integrated within a Mesoamerican world system, exploiting northern resources and channeling desirable goods into central and southern Mexican polities.

Other recent interpretations have turned attention away from foreign influences, and instead focus upon largely indigenous developments. McGuire (1980, 1986, 1989), for example, views Paquimé as the center of a "prestige economy" based upon the production of socio-economic items destined for elites in distant Mogollon and Anasazi regions.

149

Alternatively, Minnis (1984, 1988, 1989) questions the role of an exchange economy, and suggests a "peer polity" model (following Renfrew and Cherry 1986) that describes sociopolitical organizations as developing through competition, feedback, and mutual interaction. Both McGuire and Minnis reflect current directions in analyses designed to explicate regional organization and levels of complexity exhibited within, and between, the various prehistoric settlement systems subsumed under the rubric of "interaction spheres" (Minnis 1984), "provinces" (Riley 1987; Plog 1979), "regional systems" (Wilcox and Shenk 1977; Minnis and Whalen 1990; Whalen and Minnis 1999), or "alliances" (Cordell and Plog 1979; Plog 1984; Upham 1982). These investigations focus upon concepts of integration and interaction between social, economic, and political dimensions of organization. The degree of differentiation and integration is interpreted as indicating levels of cultural complexity manifest within a particular group or system.

The issues of integration and interaction are addressed with regard to the Animas phase assemblages and their relationship to Paquimé. The Animas phase data are derived from analyses of excavated materials from the Joyce Well site.

THE ANIMAS PHASE IN TIME AND SPACE

The Animas phase settlements encompass an area occasionally referred to as the "international four corners," defined by the political boundaries of Arizona, New Mexico, Sonora, and Chihuahua. Here, the northernmost reaches of the Sierra Madre Occidental extend into southwestern New Mexico from Chihuahua, Mexico, and are part of the Basin-and-Range Province. This area is one of north-south trending mountain ranges and associated valleys, with wooded uplands, wide grass-covered river valleys, and broad expanses of internal drainage systems or *playas*.

Four adjacent playas, the Animas, San Luis, Playas, and Hachita valleys comprise the "core area" of the Animas phase complex. This complex is characterized by large pueblos of coursed adobe situated around one or more plazas, rooms with regular posthole patterns and small circular firepits, scooped metates, and a ceramic assemblage that includes several plain and decorated Chihuahuan wares (McCluney 1965b, this volume). Animas phase sites have also been reported from the San Bernardino, San Simon, and Sulphur Spring valleys in southeastern Arizona (Douglas 1987; Johnson and Thompson 1963; Lekson and Klinger 1973; Sauer and Brand 1930). In Mexico, Animas phase sites have been recorded in the northeasternmost corner of Sonora, and in the northwestern corner of Chihuahua (Brand 1943; Sayles 1936a). Although settlement data for the Animas phase are few, a hierarchical settlement pattern is suggested, with large sites (100 to 500 rooms) separated by distances of approximately 16 km, with two to four smaller habitation sites located within the territory of each large settlement (DeAtley 1980:30).

Early descriptions of the Animas phase were based upon survey data compiled by Gladwin and Gladwin (1934) and Sayles (1936), and through the excavation of Pendleton Ruin, in Hidalgo County, New Mexico (Kidder et al. 1949). When Kidder and the Cosgroves initiated excavations at Pendleton Ruin in 1933 they considered it to be an "outpost" of Paquimé. However, at the conclusion of their work they suggested that the absence of certain traits held by them to be characteristic of Paquimé (i.e., subfloor inhumations, collared postholes, platform hearths, keyhole doorways) indicated a peripheral development for which they proposed the term "Animas"; a term previously advanced by Gladwin and Gladwin (1934) and Sayles (1936a), both of whom applied the term to sites adjacent to Paquimé. Though subsequent work (McCluney 1965b, this volume) in southwestern New Mexico has established the presence of those traits originally thought to be absent from Animas phase assemblages, the concept of the Animas phase as a peripheral manifestation has continued to predominate archaeological interpretations (e.g., DeAtley 1980; DeAtley and Findlow 1982; Minnis 1984, 1988; Minnis and Whalen 1990; Whalen and Minnis 1996, 1999).

The temporal placement of the Animas phase, like that of its spatial relationship, is intricately linked to Paquimé. Initially, researchers relied most heavily upon the ceramic cross-dating of trade wares, among which St. Johns Polychrome, Gila Polychrome, Tonto Polychrome, Pinedale Polychrome, El Paso Polychrome, and Chupadero Black-on-white figure prominently. On this basis, Brand (1935) suggested a range of A.D. 1300–1450, Kidder et al. (1949) proposed an occupation dating to the early or mid 1300s for Pendleton Ruin, and McCluney estimated the occupation of Clanton Draw between A.D. 1350 and 1375, Box Canyon between A.D. 1350 and 1380, and Joyce Well between A.D. 1250 and 1400 (McCluney 1965a; this volume). Following the excavation of Paquimé, Di Peso (1974) placed the Chihuahuan ceramic assemblage present at Animas phase sites within the Medio period, which he dated A.D. 1060–1340.

Absolute dating techniques have been sparingly applied to Animas phase sites. Radiocarbon assays were run by Isotopes, Inc. for three samples from the Joyce Well site. Although these ^{14}C samples were not mentioned by McCluney in his manuscript, the results were reported by Di Peso (1974:46) as A.D. 1620 ± 110 (I-1789), A.D. 1590 ± 100 (I-1790), and A.D. 1565 ± 110 (I-1791), and were considered by Di Peso to reflect a later Robles phase (Tardio period) occupation. The additional known radiocarbon values are reported by DeAtley (1980). Seven of these dates fall between A.D. 1060 and A.D. 1420, with an average calibrated age of A.D. 1216; the remaining two values are given as A.D. 875 and A.D. 545 (DeAtley 1980:69).

Obsidian hydration dating has also been conducted for some Animas phase settlements (DeAtley 1980). DeAtley (1980:77–80), employing the hydration rate determination for the Antelope Wells source developed by DeAtley and Findlow (1980), suggests that the Animas phase can be bracketed by a range of 3.9 microns (with a calendar date of approximately A.D. 1185) to 2.7 microns (circa A.D. 1400).

Most recently, Schaafsma and others (this volume) recovered archaeomagnetic samples from four hearths at the Joyce Well site, with the results providing three usable dates. These three dates, when combined, provided a relatively tight range of A.D. 1345–1370.

THE VIEW FROM JOYCE WELL

Those traits traditionally thought to be absent from Animas phase assemblages, specifically keyhole doorways, collared postholes, platform hearths (including the ornate "scalloped" style found at Paquimé), and subfloor burials are, in fact, common at the Joyce Well site. These features were also present at Clanton Draw and Box Canyon (McCluney 1965a). Moreover, ball courts are present at the Joyce Well site, Timberlake Ruin, and the Culberson Ranch Ruin (Ravesloot and Foster 1984; Skibo and Walker, this volume).

In summarizing the salient characteristics of the Joyce Well site, McCluney (this volume) suggests that it represents a typical Animas phase settlement, whose inhabitants were Paquimeños who constructed and occupied the site for the express purpose of ". . . instituting trade and to serve as an agricultural center for the raising of food crops for . . . Casas Grandes. . . ." McCluney further notes that the numerous grinding implements may indicate an inordinate emphasis on the processing of foodstuffs, perhaps related to the exchange economy (this volume). Citing the abundance of El Paso Polychrome, McCluney (this volume) suggests that extensive trade between the Animas phase and El Paso phase settlements was an important economic association, possibly linked to the exportation of foodstuffs ultimately destined for Paquimé.

Based on the data set outlined above, two primary goals were established for the analyses and results summarized here: (1) refinement of the Joyce Well chronological framework, and (2) the analysis of selected ceramic attributes to investigate relationships between the Joyce Well site and adjacent areas.

CHRONOLOGY

The dating of the Medio period—the flourescent era of Paquimé and the Chihuahua Mogollon—has recently become the subject of much debate. The chronology presented by Di Peso relies on his tenuous and subjective readings of tree-ring and radiocarbon dates. The beginning date of the Medio period (A.D. 1060–1340) is based on the terminal date of the

Viejo period derived from a single ^{14}C date of A.D. 1060 ± 190 from a floor with six Mimbres Black-on-white sherds in association, and on the calendrical dates derived from the outer rings of three construction timbers from which an unknown number of outer rings are missing (Di Peso et al. 1974:4:2–33). Those dates from Paquimé, which were significant in establishing the end of the Medio period, are a ^{14}C date of A.D. 1310 ± 30 taken from a pit oven, and a tree-ring date of 1338vv from a support timber. There is no reason to believe that these dates reflect either the beginning or end of the Medio period.

While LeBlanc (1980b) agrees with Di Peso on the beginning date, he suggests that the Medio period was in decline by A.D. 1300. On the other hand, Carlson (1982), Doyel (1976), Lekson (1984), and Wilcox and Shenk (1977), considering the trade wares present at Paquimé, support a beginning date of circa A.D. 1200, with a terminal date sometime in the first half of the fifteenth century. Ravesloot et al. (1995) reexamined the tree-ring dates from Paquimé and suggest A.D. 1200–1425 is a more appropriate range.

A new obsidian hydration rate was developed for the Antelope Wells obsidian source using a modified version of the induced hydration experiment developed by Michels et al. (1983).[1] This technique allows the calculation of the effective hydration temperature (EHT) through the modeling of heat flow through the soil matrix; the obsidian hydration rate for a specified temperature may be determined by solving the Arrhenius equation (a detailed description of this methodology is presented by Stevenson et al. 1989).

A histogram showing the hydration rim widths for Antelope Wells obsidian samples from Joyce Well is presented in Figure 7.1. Obsidian hydration dates were calculated for each of the artifacts with hydration rim widths less than 3.0 microns. Hydration rim measurements greater than 2.9 microns are most likely due to the scavenging and reuse of obsidian artifacts from earlier sites within the vicinity of Joyce Well. This behavior is evidenced by the fact that four of the specimens within the dating cluster have hydration bands of significantly greater width on the opposite surface of the artifact. The 2.9 to 2.2 microns cluster resulted in an age range of A.D. 1147 ± 118 years to A.D. 1537 ± 88 years.[2] If the four tail-end specimens are eliminated, the chronological range for the remaining ten artifacts is A.D. 1247 to 1330 at the 95 percent confidence interval.

Radiocarbon assays were run by Isotopes, Inc. for the three samples recovered from the site. As was noted earlier, these samples were not mentioned by McCluney, but the results were reported by Di Peso (1974:46) as A.D. 1620 ± 110 (I-1789), A.D. 1590 ± 100 (I-1790), and A.D. 1565 ± 110 (I-1791). While the exact context of these samples is not stated, Di Peso indicates that these dates were derived from corn (Di Peso et al. 1974, fig. 327-4). However, when dating corn, a C-3 grass absorption correction must be employed to account for the differential

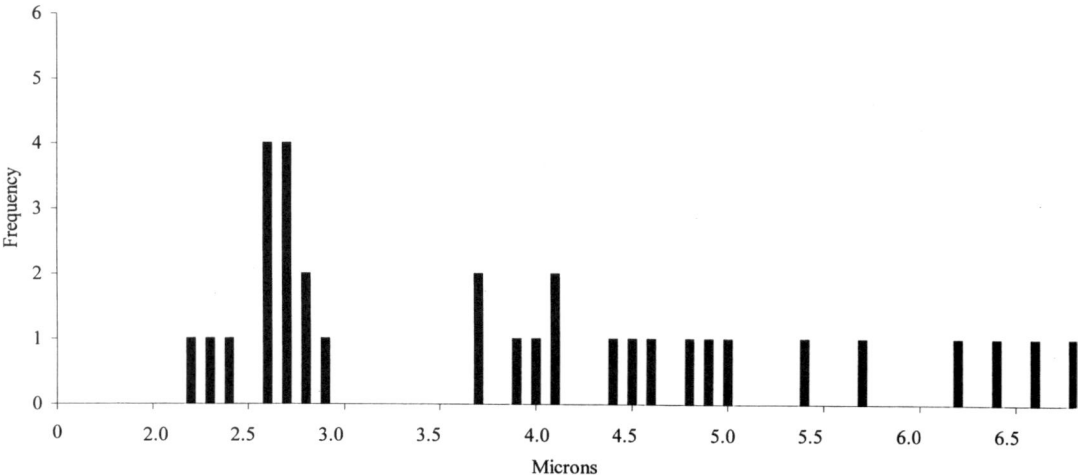

FIGURE 7.1.
Histogram of Joyce
Well hydration rim
widths (after Steven-
son et al. 1989).

absorption of radiocarbon isotopes: corn absorbs less of the lighter car-
bon isotopes and more of the heavier isotopes than is average for wood
(e.g., Lowdon 1969). Radiocarbon analysts currently suggest that an
additional 200 years be added to compensate for this discrepancy (Mur-
ray Tamers, Beta Analytic, Inc., pers. comm. 1986). The corrected dates
for Joyce Well are presented in Table 7.1.

These dates can be averaged following the algorithm presented by
Long and Rippeteau (1974) and, when weighted for the differences in
standard deviation, the average date for the three determinations is A.D.
1390 ± 60.

Finally, the El Paso Polychrome rim sherds from the Joyce Well site
were compared with the seriation suggested by Whalen (1978). The rim
forms of the El Paso Polychrome vessels, all apparently from large jars,
uniformly reveal the tight constriction below the lip and squared rim
forms characteristic of the late end of the seriation.

The results of these chronometric analyses summarized here are in
relative agreement both with one another, and with the range of occupation suggested by McCluney and others on the basis of the nonlocal
ceramics present at the Joyce Well site. Collectively, these methods indicate an occupational span centered around the middle 1300s, with a
range of approximately A.D. 1200 to 1450.

Based on Di Peso's interpretation of the uncorrected radiocarbon values from Joyce Well, he considered this site to be one of only two excavated sites indicative of a Tardio period (post-collapse) refuge population from Casas Grandes (1974:3:836). Phillips and Carpenter (1999)
question Di Peso's definition of the Robles phase, and suggest that these
sites actually date to the Medio period; this new dating of the Joyce
Well site supports a greater duration for the Medio period and casts further doubt regarding the viability of the Robles phase.

TABLE 7.1. Corrected Radiocarbon Age Determinations
for the Joyce Well Site

LAB NUMBER	¹⁴C AGE B.P.	C-3 CORRECTION	CALENDAR DATE
I-1789	385 ± 110	585 ± 110	A.D. 1420
I-1790	355 ± 100	555 ± 100	A.D. 1390
I-1791	300 ± 90	500 ± 90	A.D. 1365

REGIONAL ORGANIZATION

In the discussion of Animas phase characteristics it was noted that the concept of a peripheral, frontier phenomenon was advanced by Kidder et al. (1949) following the excavation of the Pendleton Ruin. However, McCluney's subsequent work at Box Canyon, Clanton Draw, and the Joyce Well site established the presence of those traits originally thought to be absent from the Animas phase assemblages, indicating that this basis for the Animas phase distinction is no longer tenable. Nevertheless, the concept of the Animas phase as a peripheral manifestation remains prevalent among recent interpretations (e.g., DeAtley 1980; DeAtley and Findlow 1982; Minnis 1984; Minnis and Whalen 1990; Whalen and Minnis 1999).

For example, DeAtley (1980), and DeAtley and Findlow (1982), compared site size, location, and the distribution of stylistic elements of Ramos Polychrome within various catchment systems in Hidalgo County, New Mexico. They came to the conclusion that the Animas phase settlements represented a frontier of Paquimé

> . . . settlements were not integrated into the Casas Grandes system as a unit, and . . . only a few of the villages formally participated in the Casas Grandes exchange network. For these villages, ties with Casas Grandes were as important as local Animas connections; nonetheless, it must be remembered that in neither case was the association very strong. (DeAtley and Findlow 1982:277)

However, their analysis of Ramos Polychrome is problematic. The assumption that intensity of interaction can be measured by stylistic similarity alone has been soundly criticized (Plog 1980:10–11). The results can also be questioned for the authors' assumption that all Ramos Polychrome in their sample reflect local manufacture at all sites within the study area (cf. Woosley and Olinger 1993).

An integral aspect of frontier models is the implied exploitation by the core as stimulus for colonization. Both Antelope Wells obsidian and Hachita turquoise have been suggested as a possible motive in colonizing the Animas region (Di Peso 1974:3:206–208; DeAtley 1980:35–36). The proximity of the Joyce Well site to the Antelope Wells obsidian source, a distance of 4 km, suggests that the site was well situated for exploiting this resource, but no evidence of specialized production has been found. Though settlement data for the region are limited, the Animas phase sites that have been recorded reflect a rather wide distribution that cannot be accounted for in relation to the restricted distribution of local turquoise and obsidian resources (cf. Northrop 1959).

Minnis (1984) has presented an argument in substantial agreement with that of DeAtley and Findlow. According to Minnis, economic interaction and integration refer to "relationships in the production, distribution, and consumption of goods and services" (1984:183). In this regard, Minnis focuses upon the production and distribution of four classes of goods—shell, copper, turquoise, and macaws—that is, those goods presumed by Di Peso to have been manufactured at Casas Grandes for exchange. The absence, or relative paucity, of these particular items recovered from the Animas region prompts the suggestion that (1) this area was at best only loosely integrated within the Paquimé system, and (2) that Di Peso may have greatly exaggerated the role of Paquimé as a center of production for exchange (Minnis 1984: 183–186). Minnis (1984:186) suggests that these may reflect the hoarding of valuable resources by the Paquimé elite.

Though Minnis is likely correct in questioning Di Peso's interpretation of Paquimé as a Mesoamerican port-of-trade, there is some evidence indicating the processing of copper ore, the working of raw shell materials, and the raising of macaws and turkeys. Approximately 39.5 kg of copper, including raw ore, sprue, and finished goods (bells, crotals, tinklers, pendants, an axehead, and a back-shield), along with 3,907,597 pieces of shell weighing over 1,300 kg, and the remains of 503 macaws (Ara sp.) and 290 turkeys (Meleagris gallopavo) associated with formal breeding areas were recovered from Paquimé (Di Peso et al. 1974:8:170ff). Vargas (1995) has recently questioned the assumption that the bulk of the copper objects from Paquimé was manufactured there, and provides a strong argument for assigning them to trade items from West Mexico. With the possible exception of turkeys, neither Ravesloot's (1984, 1988) mortuary data nor Di Peso's (1974) spatial distributions indicate local consumption or a significant correlation between elites and the classes of goods examined by Minnis.

The limited occurrence of these "elite" goods among regional settlements associated with Paquimé does not presuppose minimal integration; in fact, the opposite may be indicated. That certain "high-status" goods functioned in establishing or maintaining long-distance politico-economic ties between elites and their communities within the Greater Southwest has been suggested (McGuire 1989; Plog et. al. 1982; Upham 1982; Upham et al. 1981). In plotting the distribution curve for copper bells and crotals, Plog et al. (1982) found that the greatest concentration occurs between 440 and 832 km from the presumed source at Paquimé, while noting a marked reduction in quantity at sites nearby. A similar distribution pattern has been suggested by McGuire (1989) for the Paquimé macaws.

It may be that the relative paucity of these particular goods cited by Minnis may indicate the lack of any need to reify and strengthen ties through the transfer of commodities with sites culturally associated with Paquimé and the Chihuahua Mogollon. The absence of these artifacts from sites immediately adjacent to Paquimé (which, following Minnis and Whalen 1990, are presumably integrated within the Paquimé regional system) would not indicate that distant Mogollon or Anasazi settlements reflect a greater degree of interaction and integration with regard to Paquimé.

Additional problems ensue when comparisons are drawn directly from Paquimé with little or no attention paid to both issues of scale and regional variability: Paquimé, however impressive, is but one site within a widespread system (e.g., Brand 1943, Di Peso et al. 1974). For example, LeBlanc (1980b) assumes that Paquimé collapsed circa A.D. 1300 because Chihuahua Mogollon materials cease to occur in the Mimbres Valley at this time. Clearly, what LeBlanc is dating is the apparent termination of this region's participation in the Paquimé regional system, and not the collapse of Paquimé itself. While discussions of interaction abound, little effort has been expended in breaking down interaction into its constituent parts; political, economic, social, or symbolic interaction may be expressed in markedly different degrees within a given region. The concept of a regional system implies an inherent regional variability, not only in terms of resources, but in local responses to interaction, including the restructuring of local traditions, and, perhaps, resistance and conflict.

INTERACTION AND INTEGRATION: THE CERAMIC DATA

In order to address questions concerning the nature of interaction and integration evidenced between the Joyce Well site and Paquimé, analyses of El Paso Polychrome, Jornada Brown, and Ramos Polychrome were undertaken. The petrographic analysis of El Paso Polychrome and what was termed by McCluney as "Jornada Brown" (undifferentiated brown ware) was designed to test the hypothesis that these Jornada Brown sherds could perhaps be the undecorated portion of El Paso Polychrome vessels. Further petrographic analyses were conducted on Ramos Polychrome sherds to determine local vs. nonlocal manufacture.

Thin-section petrography of 21 sherds from the Joyce Well site resulted in the identification of seven distinct temper groups, all derived from crushed or decomposed igneous sources (Table 7.2). Temper Group 1 was comprised entirely of Ramos Polychrome, Temper Group 2 included Ramos Polychrome, El Paso Polychrome, and Jornada Brown samples. Temper Group 3 was associated with both Ramos Polychrome and El Paso Polychrome. Temper Group 4 included Ramos

TABLE 7.2. Temper Groups for Selected Joyce Well Sherds

ARTIFACT NO.	POTTERY TYPE	TEMPER GROUP
2035	Ramos Polychrome	1
2417-11	Ramos Polychrome	1
762	Ramos Polychrome	1
2479	Ramos Polychrome	1
563	Ramos Polychrome	1
971	Ramos Polychrome	1
627	Ramos Polychrome	1
598	Ramos Polychrome	1
789	Jornada Brown	2
2054	Jornada Brown	2
2225-5	Jornada Brown	2
627-8	Ramos Polychrome	2
723	El Paso Polychrome	2
2417	Ramos Polychrome	3
563-2	El Paso Polychrome	3
2456	El Paso Polychrome	3
2417-16	Ramos Polychrome	4
704	Jornada Brown	4
987	Ramos Polychrome	5
2225-12	Ramos Polychrome	6
2226	Ramos Polychrome	7

Polychrome and Jornada Brown. Temper Groups 5, 6, and 7 were represented solely by Ramos Polychrome.

Lacking sufficient samples from the vicinity of the Joyce Well site it is not possible to draw a one-to-one correspondence between temper groups and geological sources. Nevertheless, some general statements can be made concerning the relationship between the ceramic temper groups and the local geology to suggest that some, if not all, of the pottery sampled was made at the Joyce Well site or in the immediate vicinity.

The vitreous and often recrystallized matrix of Temper Groups 1 and 2 is characteristic of welded tuffs represented by several different formations that outcrop within a short distance of the site (Zeller and Alper 1965; Zeller 1962). Furthermore, there is considerable overlap in terms of mineral suites between Temper Groups 1 and 2 and the Park and Gillespie Tuff respectively, both of which outcrop within 4 km of the site. This association must be considered tentative without further analyses.

In the case of Temper Group 3, the association is somewhat more secure. The Center Peak latite is the only formation with a sufficiently high percentage of green hornblende to compare with that of the ceramic samples (Zeller and Alper 1965:52–54). This formation occurs approximately 16 km north of the Joyce Well site. Temper Groups 4

TABLE 7.3. Undecorated/Decorated Ratios for El Paso Polychrome

VESSEL NO.	PERCENTAGE OF UNDECORATED ZONE	PERCENTAGE OF DECORATED ZONE	RATIO
A81.3.34	49	51	0.9:1
A81.3.35	54	46	1.2:1
A47.22.16	55	45	1.2:1
None	57	43	1.3:1
None	57	43	1.3:1
A36.84.2	60	40	1.5:1
A81.3.21	62	38	1.6:1
A61.3.31	65	35	1.9:1
A81.3.25	66	34	1.9:1
None	66	34	1.9:1
A81.3.29	68	32	2.1:1
No number	68	32	2.1:1
A80.1.186	69	31	2.2:1
A75.7.197	70	30	2.3:1
A81.3.55	70	30	2.4:1
A81.1.56	71	29	2.4:1
A81.3.28	77	23	3.3:1
A81.3.33	79	21	3.7:1
A81.3.18	80	20	4.0:1
X	65.42	34.58	2.0:1

through 7 are most likely derived from processing a conglomerate. The OK Bar Conglomerate outcrops at the Joyce Well site (Zeller 1962). This formation consists of an unsorted breccia derived primarily from the Park and Gillespie tuffs (Zeller and Alper 1965:58).

There is a reasonably strong correlation between the temper types characterizing El Paso Polychrome and Jornada Brown at the site. While the evidence is considered tentative, the possibility that these two types are, in fact, a single type, El Paso Polychrome, presents a provocative implication for the Joyce Well ceramic assemblage. When the 5,054 sherds of Jornada Brown are considered together with the 939 sherds identified as El Paso Polychrome, the combined total comprises 60 percent of the pottery recovered from the Joyce Well site.

This possibility is further enhanced by an examination of the statistical relationship between undecorated and decorated portions of El Paso Polychrome jars. Typically, El Paso Polychrome ollas are globular in form, with the zone of decoration restricted to the upper portion of the vessel. Nineteen El Paso Polychrome ollas in the University of Texas, El Paso Centennial Museum collections were examined in order to determine the relationship between the undecorated and decorated zones. The results (Table 7.3) indicate that the undecorated area of these ollas varies from 49 percent to 80 percent of the total surface area, with a mean of slightly more than 65 percent. The ratio of undecorated sherds to decorated sherds from a given El Paso Polychrome olla can be expected to range from 0.9:1 to 4:1, with an average ratio of approximately 2:1.

It should be noted that this ratio refers to the zone of decoration, and not to the actual percentage of painted to unpainted space. Accounting for the unpainted areas within the zone of decoration could produce a significantly higher ratio. Thus, the 5:1 ratio expressed between Jornada Brown and El Paso Polychrome at the Joyce Well site is considered to be consistent with these figures.

Also supporting the probability of this relationship, but with far more important ramifications, is the tentative suggestion that at least some of the El Paso Polychrome may have been manufactured at the Joyce Well site. El Paso Polychrome is generally considered indigenous to the Jornada Mogollon region, where it is the ubiquitous decorated ware of the El Paso phase. The widespread distribution of El Paso Polychrome throughout northwestern Chihuahua and its particular abundance (17,068 sherds) at Paquimé prompted Di Peso to characterize this type as "the tin-can" of Paquimé (Di Peso et al. 1974:8:141), an observation that today seems astute. Wherever the locations of manufacture might prove to be, El Paso Polychrome is, by all appearances, a trademark of the *internal exchange* network within the Paquimé-Chihuahua Mogollon-Black Mountain-El Paso regional system.

As McCluney (this volume) suspected, it appears that Ramos Polychrome was manufactured at, or in the vicinity of, the Joyce Well site, where it comprises 97 percent of the Chihuahua polychromes present. Although archaeologists tend to accept the Chihuahua polychromes as indigenous to Medio period Paquimé (e.g., Di Peso et al. 1974:6:183 ff), there is evidence indicating that the distributions of Babicora, Carretas, Huérigos, Ramos, and Villa Ahumada polychromes have strong regional associations (Brand 1935, 1943): Ramos Polychrome is the predominant Medio period polychrome within an approximate 50 km radius of Paquimé; Babicora Polychrome predominates to the south of Paquimé; Villa Ahumada Polychrome is the predominant Chihuahuan polychrome to the east; and Huérigos and Carretas polychromes are most prevalent in the region northwest of Paquimé (Fig. 7.2).[3]

That Ramos Polychrome is the most prevalent Chihuahua polychrome at Joyce Well, which is separated from Paquimé by a distance of 120 km and an intervening area characterized by Huérigos and Carretas polychromes, suggests possible linkages between the two sites. All other Chihuahua polychromes combined account for less than 1 percent of the Joyce Well ceramic assemblage. Douglas's (1992) recent study of nonlocal ceramics at Paquimé concludes that pottery did not function as a prestige good, and was relatively unimportant in the regional exchange system. This conclusion seemingly holds for the Chihuahuan polychromes as well.

FIGURE 7.2. Regional distribution of Chihuahuan polychromes and associated ball courts.

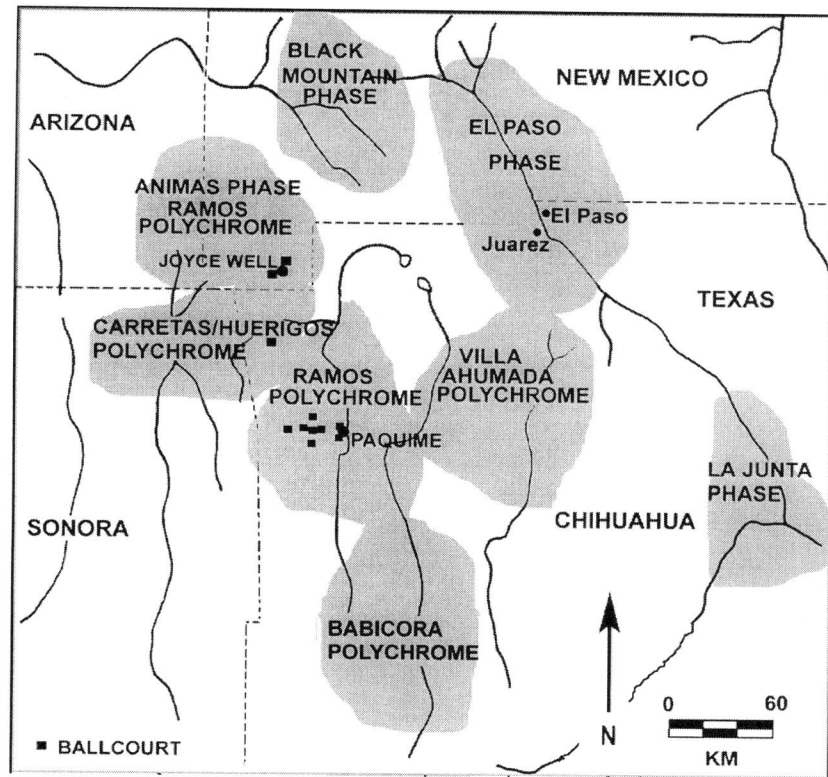

DISCUSSION

The data from Joyce Well allow new views to be explored. The presence of a ball court is suggestive of ritual intercommunity integration, and implies the existence of an organizational hierarchy (Skibo and Walker, this volume; Wilcox 1991; Wilcox and Sternberg 1983). Chihuahua Mogollon ball courts, recently thought to be associated only with Paquimé (Minnis 1984:189), have since been observed at several additional sites (Minnis and Whalen 1990; Naylor 1985).

When the known distribution of ball courts is plotted with the Medio period polychromes, a correlation emerges (Fig. 7.2). With one possible exception, ball courts appear to be associated with the occurrence of Ramos Polychrome as the predominant Chihuahua polychrome. This association suggests that the ritual behavior associated with the ball courts was an important institutional-symbolic mechanism in integrating those communities most directly associated with Paquimé (e.g., McGuire 1986; Minnis and Whalen 1990; Skibo and Walker [this volume]; Wilcox 1991; Wilcox and Sternberg 1983). Whalen and Minnis (1996:743), nevertheless, caution that the presence of ball courts are not necessarily indicative of "tight integration" within the Paquimé regional system and, along with Douglas (1995), argue that the Animas phase components do not reflect a well-integrated, dependent periphery.

However, regional variability within the Chihuahua Mogollon region, indicated by the differential distribution of polychrome traditions, can be discerned in the Ramos (Paquimé), Babicora, Carretas-Huérigos, and Villa Ahumada subregions. Minnis and Whalen (1990; Whalen and Minnis 1996, 1999) have suggested that the Paquimé regional system encompassed an area within a 30 to 40 km radius. This area roughly corresponds with what I identify as the Ramos subregion. The evidence from the Joyce Well site suggests that at least this local expression of the Animas phase complex is associated with the Ramos subregion, and is likely related to the growth of Paquimé as a regional center.

In contrast, the Black Mountain phase (also occasionally referred to as "Animas phase" [e.g., Ravesloot 1979]) in the Mimbres Valley and El Paso phase of the Lower Rio Grande region witness a marked departure from their traditional developments in the adoption of a number of Chihuahuan characteristics, manifest in the construction of compound puddled-adobe architecture, platform hearths, collared postholes, and the manufacture of Chihuahuan-style pottery and ground stone (LeBlanc and Nelson 1976; Lehmer 1948; Ravesloot 1979; Shafer 1999). Changes in Black Mountain phase settlement patterns, which exhibit a trend towards more southerly locations and lower elevations, may reflect an attempt to facilitate connections with Paquimé (Ravesloot 1979:87–88). The expression of Chihuahuan traits in these regions is interpreted as evidence for participation in an expanding socio-politico-economic system centered at Paquimé. A similar phenomenon may have occurred during the La Junta phase (A.D. 1200–1400) in the Junta de los Rios region of the Rio Grande, where high percentages of El Paso and Ramos Polychromes occur in association with a shift to puddled-adobe architecture (Kelley 1948, 1951; Riley 1987).

Unfortunately, the exact nature of these relationships remains unknown. A cursory examination of the commodities recovered at Paquimé reveals that the majority were procured from within the proposed Paquimé-Chihuahua Mogollon-Black Mountain-El Paso regional system (see Di Peso et al. 1974:8:141–192). The dispersed number of commodities within household and storage contexts can, perhaps, be interpreted as reflecting market acquisition, as defined by Pryor (1977:106). In this system, both subsistence and exotic goods may be obtained through barter in both goods and services. LeBlanc (1975) has suggested that labor may have been a valuable Mimbres commodity exchanged within the Paquimé system.

Yet, the rapid transformation of the Black Mountain phase Mimbres and El Paso phase Jornada regions into "ersatz Chihuahua culture" is indicative of a nonlinear trajectory that cannot easily be explained solely as the result of exchange relations. Participation in the Paquimé regional system, in these cases, seemingly implies sociopolitical ramifications above and beyond that necessitated by economic relations.

Though by no means unequivocal, this interpretation is complementary to the suggestion that power and prestige at Paquimé may have been grounded, in part, on warfare (Douglas 1992:20; Ravesloot 1988: 75–76; Ravesloot and Spoerl 1987).

Additionally, there are those regions with which Paquimeños may have maintained long-distance relations through the specialized production and distribution of rare and desirable commodities. The macaw, copper, and possibly shell, industries appear to be primarily directed toward long-distance distribution. Some view the distribution of these goods as evidence of alliances between elites (e.g., Plog et al. 1982; Upham 1982), whereas McGuire (1989) suggests that these items are associated with the development of the katsina cult. In either case, the prestige goods economy appears to be explicitly linked to groups in the northern portion of the Greater Southwest; aside from the probable exception of copper bells (e.g., Vargas 1995), economic interaction with Mesoamerica is not readily indicated.

The prestige goods economy likely reflects an organization distinct from that of the regional system. According to Wolf (1982:77–110), prestige goods economies are identified with kin-based modes of production. As described, the Paquimé regional system may reflect the transcendence of a strictly kin-based power structure. Thus, the distribution of variables associated with the prestige goods economy may not be a suitable measure for interaction and integration with regard to the regional system. Instead, I suggest that Ramos Polychrome and ball courts are more likely indicative of intracultural integration with Paquimé than are commodities such as turquoise, copper objects, shell, or macaws. Intercultural interaction and integration, as reflected by the Black Mountain phase of the Mimbres and the El Paso phase of the Jornada Mogollon (and, perhaps, the Junta de los Rios region), can be seen in the adoption of Chihuahua Mogollon traits and changes in settlement patterns as well as in the distribution of commodities. The proliferation of religious-ceremonial features constructed at Paquimé during the Medio period, evident in the construction of 2 ball courts, 18 platform mounds, and other specialized structures like the House of the Well, may be interpreted as evidence of need for integrative mechanisms within a society becoming increasingly differentiated (e.g., Yoffee 1979:28).

This perspective, like that proposed by Minnis (1984, 1988, 1989), suggests that the growth trajectory associated with Paquimé and the surrounding regions can be explained in terms of local processes. However, the degree of differentiation evident within the Paquimé regional system and the likelihood of both market and prestige goods economies indicate systems of interaction and integration that cannot be simply defined by concentric rings of decreasing spheres of interaction radiating out from Paquimé. In turn, this intimates a level of complexity which, as yet, awaits adequate illumination.

CONCLUSIONS

The interpretation of Animas phase settlements as a peripheral, frontier phenomenon at best loosely linked to the cultural developments at Paquimé is attributed to the measure of inappropriate variables. It has been shown that the basis for the original distinction of the Animas phase by Kidder et al. (1949) is unfounded. Di Peso's interpretation of the Joyce Well site as a Robles phase refuge for peoples having fled a sacked and destroyed Paquimé is not supported.

The Animas phase settlements are recognized as Medio period Chihuahua Mogollon sites whose major distinction as a complex can be attributed to the international border. The Medio period Chihuahua Mogollon are presented as comprising at least four subregions that can be differentiated on the basis of polychrome traditions; the Ramos (Paquimé), Carretas-Huérigos, Villa Ahumada, and Babicora regions have been proposed. The local manufacture of Ramos Polychrome and the presence of a ball court suggest that the Joyce Well site is closely identified with Paquimé.

At least two different spheres of interaction can be defined; the specialized production of commodities for long-distance exchange likely reflects a different economic organization from that associated with the Paquimé-Chihuahua Mogollon-Black Mountain-El Paso regional system. I have suggested that the analysis of interaction consider the various types of interaction, including social, political, economic, and symbolic dimensions, with particular regard to regional variation. This is especially important if we are to specify what integration and interaction mean, not only in terms of socio-politico-economic organization, but in the restructuring of local traditions, as is apparently the case for the Black Mountain and El Paso phase manifestations.

There is currently no evidence to suggest that the Joyce Well site was established in order to facilitate the extraction of local mineral resources. The high percentage of El Paso Polychrome may not indicate direct transfers with the Lower Rio Grande region, but local manufacture associated with participation in a Paquimé-dominated regional system that included the Black Mountain and El Paso phase areas.

The seemingly excessive amount of ground stone along with the abundant storerooms of maize may yet support McCluney's original assessment of an economy at least partially based on the export of foodstuffs to Paquimé. In this case, we are presented with the possibility that a region some 120 km north of Paquimé was colonized by peoples from Paquimé or its immediate vicinity in an effort to expand the production of agricultural surplus.

Another possible alternative is that Animas phase sites such as Joyce Well may have been established in order to circumvent the Carretas/Huérigos subregion, and facilitate interaction with regions to the north and, perhaps more importantly, to the west. The majority of the marine shell recovered from Paquimé, including *Glycymeris, Conus, Oliva, Olivella,* and *Nassarius,* was identified as originating from the Sea of Cortez (Di Peso et al. 1974:4:401). Ongoing investigations at La Playa, in northern Sonora, indicate that this region has been actively involved in the mass production of shell ornaments since at least the Cienega phase of the Late Archaic/Early Agriculture period (800 B.C. to A.D. 200) (Carpenter et al. 1999; Carpenter and Villalpando 2001), and recent excavations at the adjacent Cerro Trincheras demonstrates that Ramos Polychrome was the predominant decorated pottery type at the site between approximately A.D. 1300 and 1450 (McGuire et al. 1999). In either scenario, the Animas phase sites may constitute a colonial "frontier," albeit one that was apparently well integrated with regard to Paquimé.

ACKNOWLEDGMENTS

This chapter is a revised version of a paper originally submitted in partial fulfillment of the Master's degree in the Department of Anthropology, New Mexico State University, and has benefitted from constructive comments from a number of individuals, including David Batcho, Paul Fish, Michelle Hegmon, Jonathan Mabry, Paul Minnis, Tom Naylor, J. Jefferson Reid, Jim Skibo, Ed Staski, and Norm Yoffee. Of course, any leaps of faith or errors in logic are mine alone. The obsidian hydration analysis was conducted by Christopher Stevenson, and the ceramic petrography was performed by David Hill; these analyses were made possible through a New Mexico State University Arts and Sciences Council Research Grant. The University of Texas, El Paso Centennial Museum graciously provided access to their El Paso Polychrome collection. A special note of thanks is due Eugene McCluney for his support and willingness to answer my many questions. The figures were drafted by Guadalupe Sanchez. Tom Naylor's passing has left a tremendous void among researchers in this region; his love and knowledge of northern Chihuahua was unexcelled, and he is greatly missed.

NOTES

1. The Michels et al. (1983) experiment is essentially a dissolution experiment where the surface of the glass begins to dissolve as the hydration rim is forming. It is not visually noticeable until higher temperatures (180° C) and longer reaction periods (10 days) are reached, but occurs at all of the temperatures utilized in the experiment (cf. Stevenson et al. 1989; White 1986). As a result, the hydrated layer thickness does not represent the true hydration history and the calculations made from the data are misleading. We (Stevenson et al. 1989:195–196) identified three potential sources of error and modified the experiment accordingly. First, the addition of silica gel to the solution effectively controlled surface dissolution

promoted by deionized water. Second, a higher solution volume to sample surface was used in the rate determination process in order to reduce pH$^{\prime}$ levels, as pH levels above 9 have been shown to promote dissolution. Third, longer experimental runs were conducted in order to minimize error introduced by optical measurement techniques.

2. It should be noted that the calculated age of the hydration rim widths differs substantially from those presented by DeAtley (1980:77–80). Briefly, the hydration rate determination developed by DeAtley and Findlow (1980) was established by correlating hydration rim widths with radiocarbon dates along a regression curve (Pearson's r). Based on this technique, DeAtley (1980:77–80) suggests that the Animas phase can be bracketed by a range of 3.9 microns (with an associated calendar date of approximately A.D. 1185) to 2.7 microns (approximately A.D. 1400). Unlike the induced hydration experiment, this technique does not control for the effective hydration temperature, which is influenced by a number of factors, including exposure and surface and subsurface thermal diffusivity.

3. Carretas and Huērigos polychromes are identical except for the white slip applied to the latter (cf. Rinaldo in Di Peso et al. 1974:6:243). This relationship is further evidenced by numerous bowls with a Carretas Polychrome exterior and a Huērigos Polychrome interior. However, some spatial differentiation is observed, with Carretas predominant at the southern end of the province and Huērigos more evident to the north.

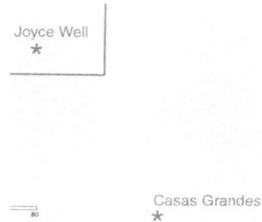

EIGHT

Joyce Well and the Casas Grandes Religious Interaction Sphere

William H. Walker and James M. Skibo

Joyce Well has played a pivotal role in the archaeology of the Chihuahuan culture horizon and continues to hold critical information for understanding ritual variability in the international four corners of the American Southwest (Sonora, Chihuahua, Arizona, and New Mexico). More than thirty years ago Caldwell (1964:142) posed the question, when "we consider the origins of civilizations . . . shall we find religion based interaction spheres?" A religious interaction sphere is a region comprised of autonomous groups whose common ritual activities result in homogenous assemblages of certain classes of material culture (Caldwell 1964). The Chihuahua culture horizon (A.D. 1200–1450) appears to have been such a religious interaction sphere centered on the large town of Paquimé (Ravesloot 1979; C. Schaafsma 1995; P. Schaafsma 1997).

Communities in this regional system, including Joyce Well, possessed similar artifacts (e.g., shared polychrome-style ceramics, scooped metates), architecture (e.g., adobe pueblos, T-shaped doors, platforms hearths,) and rock art (e.g., feathered serpents, cartouches) indicative of a reservoir of shared beliefs and ritual practices (e.g., Di Peso 1974; Ravesloot 1979; P. Schaafsma 1998). The organization of religious interaction in the region, however, remains murky because, until recently, the majority of research has focused on Casas Grandes. It is not clear how ritual activity and leadership was organized in the various outlying communities of the region. Although Casas Grandes is relatively well known (Di Peso 1974; Di Peso et al. 1974), the majority of sites remain untested or unreported (Schaafsma and Riley 1999:9;

Whalen and Minnis 1999:36). Until more detailed excavation and pub-lication occurs it will be difficult to understand the social, religious, and economic processes that shaped the Casas Grandes world between A.D. 1200 and 1400.

In addition to possessing regionwide traits, Joyce Well also has a ball court. The archaeological record in the Bootheel more closely links this region to Casas Grandes and suggests the possibility of some form of prehistoric religious integration (Ravesloot 1979:89; P. Schaafsma 1997; Wilcox 1995), perhaps involving pilgrimage activities (Fish and Fish 1999:23–24). This integration does not appear to involve a regionally powerful priesthood but instead local communities of various sizes with varying forms of religious leadership. The peoples of the Bootheel em-ployed similar icons and ritual buildings (plazas and ball courts) in their own local ritual tradition that never attained the complexity of Casas Grandes.

The depiction of horned snake imagery, parrots, and abstract icons, and the use of stylistically similar architecture and artifacts (e.g., Ramos Polychrome ceramic styles, platform hearths, ball courts, and plazas) across the interaction sphere suggest a reservoir of shared beliefs and practices at the household and community scale. The largest centers, such as Casas Grandes, would attract the most ceremonial goods. The hoarding and use of shell, parrots, copper, and Gila Polychrome pottery at Casas Grandes may simply be a product of their use at the center (Lekson 2000; Minnis 1984, 1988; Whalen and Minnis 2001:183–184). In such a scenario, these materials were manufactured, stored, used, and often ritually discarded by a ceremonial constituency of town residents and perhaps pilgrims.

Whalen and Minnis (2001) following Sebastian's (1992) study of the Chaco world suggest Casas Grandes' rise was engineered by Viejo com-munities with better access to agricultural resources. These groups along the Rio Casas Grandes used their limited surplus food to establish a prestige economy that crystallized in the short-lived fluorescence of ceremonial structures and multistory room blocks at Casas Grandes. Eventually, the costs were too high and the surplus food and labor re-quired to maintain the town dried up. On a much smaller scale we en-vision a similar process in the Animas phase pueblos. Survey evidence to date suggests that these were the first and last aggregated settlements in the Bootheel. The drive for such aggregation and its subsequent failure paralleled the rise and fall of Casas Grandes. We believe that evidence of community and household ritual power at Joyce Well, therefore, may reside in its abandonment stratigraphy. If economic production at the Joyce Well site was based on kin networks whose members voluntarily participated in the formation of a pueblo community, then the fate of the community and its ritual organization required that economic inter-ests and ritual activities follow similar paths. When those paths di-verged, the ritual integration as well as community organization would collapse. Stratigraphic evidence of abandonment activities, therefore, may contain critical information about the timing and nature of community disintegration at both sites.

Although timing of abandonment throughout the interaction sphere has not been established, preliminary data (Chapter 6) suggest that the Animas phase and the Casas Grandes core area were both abandoned in the mid to late fourteenth century. Joyce Well, like many other sites in the region including Casas Grandes, exhibits evidence of fiery destruction that has been attributed to prehistoric violence by some (Di Peso 1974; LeBlanc 1999; Ravesloot and Spoerl 1989; Turner and Turner 1998). Ongoing Southwestern archaeological research, however, suggests that this burning may have resulted from ritual activities rather than war (e.g., Walker 1998, 2002; Wilshusen 1986). This admittedly controversial hypothesis, if supported, will have profound implications for the study of the interaction sphere and the archaeology of prehistoric ritual more generally. Warfare and coercion have been central tenets of previous models depicting a politically and economically centralized polity centered at Casas Grandes. The context of burning may instead document variation in household and community integration across the interaction sphere.

Although religions are often described as ideologies or systems of beliefs, their organization also entails concrete interactions between people and artifacts that have immediate goals (e.g., healing, protection, community well-being, revenge) and result in tangible traces in the archaeological record. Ritual can, therefore, be categorized for analysis as a form of technology. As behavioral archaeologists we share a common interest with many materially oriented scholars in assessing how the organization of technologies, including ritual ones, result in variable human creations (artifacts, features, deposits) ranging from clay cooking pots to stratigraphic layers of burned roofing material containing ceremonial trash. Future work at Joyce Well and the larger interaction sphere would benefit from the study of the ritual activities shaping stratigraphic variability.

ARTIFACT LIFE HISTORIES AND STRATIGRAPHY

Cultural processes, including war and ritual, shape the life histories of artifacts from their acquisition as raw materials, manufacture, distribution, uses, and eventually deposition (Schiffer 1987). Just as ethnologists model the organization of behaviors in relation to culture, archaeologists model the organization of the structure of archaeological sites in relation to natural and cultural processes (Binford 1978, 1987:461). Beginning in the 1970s, explicit, albeit schematic, artifact life-history models of inference were initially proposed (Schiffer 1972; Sullivan 1978) to link the final stages in artifacts histories (archaeological context) with earlier stages (systemic context).

In the last decade, Southwestern archaeologists have attempted to model various stages in the life histories of objects used in local and regional cults (Adams 1991; Bradley 1996; Crown 1994; LaMotta and Schiffer 1999; Montgomery 1993; Walker 1998; Wilshusen 1986, 1988). Crown's (1994) study of the manufacture and distribution of Salado polychrome ceramics, for example, has demonstrated that there was not a regional Salado "culture," but instead a religious ideology shared by economically and politically independent Southwestern potters. In another example, Adams (1991) described the origin and spread of the prehistoric katsina cult tracing the use of katsina mask iconography, rectangular subterranean ceremonial structures (kivas), enclosed dance plazas, and innovations in communal cooking technologies (e.g., piki stone, griddles) across multiple independent Pueblo cultures. Bradley's (1996) recent study of the ritual organization of thirteenth-century Anasazi communities, such as Sand Canyon Pueblo, makes the case that after the collapse of a macroregional religious system centered at Chaco Canyon, smaller regional cults on its periphery revitalized or reused Chacoan architecture and artifacts (Mesa Verde mugs) to legitimize new cult practices. Growing archaeological interest in regional cults worldwide has promoted an interest in defining stratigraphic traces of cult practices (e.g., Freidel and Schele 1988; Grieder et al. 1988; Mock 1998).

Wilshusen's (1986) pioneering study of household and village ritual abandonment activities initiated a series of stratigraphic studies (e.g., LaMotta and Schiffer 1999; Montgomery 1993; Walker 1995a) of household and community religious practices. His studies of the strata contained in Pueblo I (A.D. 700–900) ceremonial structures documented a fundamental assumption of object life-history research: earlier stages in a structure's history, including its ritual uses, condition later stages such as its ritual abandonment through purposeful burning or burial.

Advocates of such models examine how the organization of known behaviors affects the frequencies, formal properties, associations, and spatial locations of artifacts. Such approaches to inference are particularly strong because they highlight ambiguity in archaeological stratigraphy that mandates more precise inferences of prehistoric social processes. Religions result in various combinations of ritual behaviors (e.g., funerals, site and structure abandonments, ceremonial trash discard) that in turn contribute to variability in archaeological strata. Indeed, the ethnographic study of regional cults and the archaeological study of religious interaction spheres arose in part to understand how cultic practices might more comprehensively explain distributions of artifact and architectural patterns than more hierarchical economic and political explanations (Burger 1992; Caldwell 1964; Keatinge 1981; Turner 1974; Werbner 1977). One of the more intriguing aspects of such cult organization is that it often thrives in the interstices of political and economic power (Coe 1981; Burger 1992). Ethnographically

middle-range societies possess local and regional cults that often organize the household, community settlement patterns, architecture, and
artifact assemblages that archaeologists describe as cultures, ethnic
groups, or cultural horizons (Caldwell 1964; Coe 1989; Schortman and
Urban 1987; Willey 1948). Understanding ritual stratigraphy generated
by such cults reveals important evidence of household and community
organization. It is just such local ritual activity that is at issue in the ongoing debate over prehistoric destruction in the Casas Grandes world.

A STRATIGRAPHIC MODEL

Stratigraphic approaches to ritual are controversial in part because they
often turn upon relatively ordinary evidence (trash fill, sand, roofing
materials, animal bones) not conventionally considered ritual in nature.
These deposit-oriented approaches to ritual can be expressed in a general model focused on the terminal life histories of a series of behav-
iorally meaningful stratigraphic units such as rooms, pits, plazas, mid-
dens, and wells. Among the ancient pueblo and pithouse villages of the
American Southwest, archaeologists have focused on the relationships
between pithouse and pueblo floor features (e.g., hearths, pits), floor assemblages (e.g., whole artifacts, fragmentary artifacts), and strata above
the floor including trash deposition, roof fall, windblown sand, etc. Any
one variable, such as roof burning at Joyce Well or Casas Grandes could
not distinguish accidental burning from warfare, arson, or insect exter-
mination. Consideration of this variable in conjunction with others in a
series of structures, however, would demonstrate, in a probabilistic
sense, more or less likely causes.

   If a majority of roofs in a pithouse village were burned, then this high
frequency of burning renders accident a less likely explanation. Indeed
this was the case for approximately 80 percent of Basketmaker pithouses in the American Southwest (Walker 1995b). Given this high frequency of occurrence, it is unlikely that these pithouses entered the archaeological record through accidental fires. It is also the case that most
of these houses lacked the intact household artifact inventories one
might expect from a surprise attack. When trash deposits and sealed
features are considered in relation to these strata, ritual abandonment
often becomes the best explanation. A number of Southwestern archaeologists studying burning among the Hohokam, Anasazi, and Mogollon
cultures have employed similar arguments to identify ritual burning
(Montgomery 1993; Seymour and Schiffer 1987; Walker 1998;
Wilshusen 1986). The point is that as one considers increasing amounts
of stratigraphic information, the range of possible explanations of roof
burning decreases.

   Application of such models to archaeological evidence is labor intensive and requires relatively specific stratigraphic information that may
not be available in older site reports such as those from the Casas

Grandes region. As a result, many ritually based arguments have focused on the strata of more recently excavated sites (e.g., Lightfoot 1993; Mock 1998; Montgomery 1993; Stevanovic 1997; Walker 1996, 1998; Wilshusen 1986). Nonetheless, older site reports possessing less intensive stratigraphic resolution often contain enough evidence to suggest regional patterns of ritual behavior (e.g., Bradley 1990; Hill 1995; Walker 1998).

Finding such data is exciting because they provide a material link to previously unknown social processes. Wilshusen and Ortman (1999:386–387), for example, found that the relationship between burning and abandonment of early Pueblo structures in the Dolores Valley of southwest Colorado not only supported ritual destruction over warfare, but also demonstrated regional ritual variability among Pueblo communities that had previously gone unrecognized. The burning reported from Joyce Well and other sites suggests we may find such patterns in the stratigraphy of sites in the Bootheel.

RITUAL ABANDONMENT AND THE ANIMAS PHASE

Although recording of stratigraphic data at Joyce Well and other Animas phase sites has been less than ideal, there are clues that these villages were ritually abandoned. McCluney (1965a:14) noted, for example, that hearth features at Pendleton Ruin, Clanton Draw, and Box Canyon "were [generally] completely free of residual ashes." Piecing together similarly broad descriptive statements about strata at the Box Canyon site (McCluney 1965a:23, 29, 36–37, 38), it would seem that relatively ordinary looking structures such as Room 12 emerge in a life-history analysis as a possible ritual structure. The only whole ceramic vessel recovered from this site, a shoe pot, was found at "subfloor level" in this room. This room's platform hearth, a recognized Casas Grandes trait, was free of ashes, and clusters of corn lay on the floor of the structure when it burned. After this room and most of the other rooms at the site had been abandoned, they "had been briefly used as trash depository by a later migratory people" (McCluney 1965a:23).

These stratigraphic clues provoke a series of contradictions that our life-history method can potentially resolve in future excavations. Why clean out hearths and remove the ceramic inventory of the village, but leave behind bushels of food? Why remove hearth ashes, and whole ceramics, and then burn this structure, but not leave the site (assuming that the later trash did not result from a migratory people)? Although a shoe pot is a well-designed implement for cooking (Dixon 1963, 1976), why was it buried whole beneath the floor of this particular room possessing a distinctive hearth? Was this shoe pot more than a utilitarian artifact? Did it also participate in a pan-Southwestern ritual tradition as Adams (1991) suggests in his study of the prehistoric katsina cult? The sequence of strata in this room, when considered as a causally interrelated series of life-history stages, suggests the possibility that a foundation deposit (a shoe pot) had been placed beneath the floor of this room during construction. Subsequently, after removing ash from its hearth, the structure was ritually burned, leaving corn on its floor as an offering. The burning of the room, however, was not associated with the abandonment of the site, as later deposits of trash overlay the burned roof. Instead, the abandonment of the room resulted from ritual events in the on-going life cycle of a household, and not villagewide ritual abandonment.

A number of striking traces of abandonment activity exist in McCluney's Joyce Well data (Chapter 2). A brief summary of the architectural and artifactual patterns highlights McCluney's insight that violence was not instrumentally involved in the abandonment and fiery destruction of rooms at Joyce Well. He exposed 45 rooms and 24 of these were burned. Trash deposits in several rooms above the burned layers as well as the superposition of burned floors in Room 16 indicate this burning was a series of events rather than one catastrophe. The room assemblages and features do not resemble those of rooms destroyed while still in use. Almost all doorways encountered were sealed. In most cases this involved filling the space with daub and pebbles. In some cases larger artifacts, such as manos or metates, were employed. In one instance, Room 26, a partial buffalo skull was sealed into a T-shaped doorway. The majority of hearths held little or no ash. Only a handful of whole artifacts were recovered from floor contexts. The overwhelming majority of whole artifacts were found in graves. Indeed even the few exceptional rooms such as Room 24 that contained stacked rows of burned corn or Room 31 possessing four El Paso Polychrome ollas possessed neither the diversity nor density of artifacts indicative of a catastrophic attack. Although approximately half of the rooms were burned, the remains of weight-bearing support posts and vigas are small relative to stringers and roofing grasses, suggesting that still-useable roofing materials were possibly recycled before or after the fire.

These data suggest several new questions for future research at Joyce Well.

1. What is the order and rate of room abandonment?
The movement of households within the village or out of the village would provide a measure of ritual integration and community stability. It appears that room abandonments are not simultaneous. Archaeomagnetic dating of burned rooms already suggests two periods of abandonment in the site (Chapter 6). Future excavations will continue to assess this pattern. In tandem with such study will be efforts to explain variability in burning.

2. Although all room abandonments appear to have been carefully planned, why are some burned and others not? Slightly more than half of the rooms (24 of 45) were burned. Based on our life-history model

we would hypothesize that ritual activities earlier in these structures' life histories resulted in more intensive ritual abandonments involving burning. Some rooms might have possessed ceremonial functions and others may have been set apart through funerary practices. Fourteen of the 19 burial contexts are subfloor interments in burned rooms. Room 24, for example, possessing the burned corn (described above), also contained the subfloor burial of the oldest individual recovered from the site, a male aged approximately 45 years. Four burial contexts, however, remain unexplained: three occur in unburned rooms and one in the plaza. Perhaps some of these unburned rooms were actually burned. Burning was not a research variable for McCluney and our inferences of burned rooms derive from wall, floor, and roofing descriptions. When burning was relevant for discussions of preservation, he noted it. In four of the five "unburned" rooms containing burials, roofing evidence is not discussed. Nonetheless, there are burned rooms that do not possess burials.

Perhaps some of these burned rooms had other ritual uses. We do not have enough evidence to assess whether the stratigraphic contents of these rooms differentiated them from unburned rooms. Nor can we differentiate them temporally. We hope to gather such information in the future.

3. What happened to the Joyce Well inhabitants after abandonment? The Animas phase, like that of the Medio period in Chihuahua, is the last recognized prehistoric phase in the region. Spanish documents vaguely refer to farmers and hunter-gatherers in the region at contact. Disease must have dramatically reduced these populations making way for Athabascan occupation. During the nineteenth century, the Bootheel was part of the territorial range of the Chiracahua Apache. We suspect that smaller farming hamlets and hunting and gathering sites dating to the fifteenth and sixteenth centuries may exist among the plain ware sites found in the drainages of the Playas and Animas valleys. Answering this question will require more survey as well as focused testing and dating of such sites.

CONCLUDING THOUGHTS IN HISTORICAL CONTEXT

The Bootheel of New Mexico and Joyce Well in particular hold many clues to the organization of the Chihuahua culture horizon, one of the largest but least understood interaction spheres of the ancient Southwest. Recognizing the potential importance of the Bootheel region to the study of Casas Grandes, Alfred Kidder, Hattie Cosgrove, and Burt Cosgrove (1939) excavated the Pendleton Ruin. They named the Animas phase and dated (based on ceramic cross-dating) Casas Grandes and the Bootheel pueblos to the fourteenth century. Frustrated by the complexities of adobe wall architecture and a lack of Casas Grandes traits, however, the Animas phase villages were deemed only remotely related to the Casas Grandes world.

Eugene McCluney's later studies of Clanton Draw, Box Canyon, and Joyce Well carried out during his directorship of the School of American Research radically transformed that perception. His extensive excavations revealed numerous artifactual clues and architectural features that directly tied the sites in this region to a cultural core at Casas Grandes. Joyce Well (Chapter 2, this volume) in particular possessed T-shaped and keyhole doorways, raised hearths, collared posts, and a range of Chihuahuan polychrome ceramics. His work provoked Di Peso (1974) and subsequent scholars to reconsider the relationship between Casas Grandes and the Animas phase. Based on uncalibrated radiocarbon dates, Di Peso (1974) envisioned it as the home of displaced refugees of the once powerful Casas Grandes trading center. Redating of Casas Grandes and Joyce Well (Dean and Ravesloot 1993; DeAtley 1980; Schaafsma et al., this volume), however, affirmed Kidder et al.'s (1949) original inference that the Animas phase sites were contemporary rather than later than Casas Grandes. The question remained, however, what tied Casas Grandes to the Bootheel?

This question led Whalen and Minnis (2001) to reexamine Di Peso's model of the Chihuahua culture. Their studies of the Casas Grandes data and surrounding regions did not support a hierarchical society dependent on a mercantile economy focused on Paquimé but instead a much smaller ranked society centered on Casas Grandes whose economy was organized by relations of kinship. More distant regions were conceptualized as concentric zones of decreasing social interaction. The Bootheel, however, seemed to defy the distance. As John Carpenter notes in Chapter 7, the predominance of ball courts and Ramos Polychrome distinguish it as more similar to the Casas Grandes core area than of peripheral regions.

Building on this previous work, our continuing investigation at Joyce Well has turned toward the study of ritual activity to explain the social interaction between the Animas Phase villages and Casas Grandes. We want to understand the use and significance of the ball courts (Chapter 5) as well as how locally organized ritual practices could be tied to regional patterns of religious interaction. We hypothesize that a reservoir of shared religious beliefs gives unity to the Chihuahuan culture horizon leading to trade and exchange activities focused on a ceremonial center at Casas Grandes. In our future excavations we will continue to fine-tune the regional chronology in order to place evidence of ritual activity, such as room and site abandonment, in a historical context.

REFERENCES

Adams, E. C.

1991   *The Origin and Development of the Pueblo Katsina Cult*. University of Arizona Press, Tucson.

Binford, L. R.

1978   Dimensional Analysis of Behavior and Site Structure: Learning from an Eskimo Hunting Stand. *American Antiquity* 43:330–361.

1987   Researching Ambiguity: Frames of Reference and Site Structure. In *Method and Theory for Activity Area Research: An Ethnoarchaeological Approach,* edited by S. Kent, pp. 449–512. Columbia University Press, New York.

Bradley, B. A.

1996   Pitchers to Mugs: Chacoan Revival at Sand Canyon Pueblo. *Kiva* 61:241–255.

Bradley, R.

1990   *The Passage of Arms*. Cambridge University Press, Cambridge, U.K.

Brand, D. D.

1935   The Distribution of Pottery Types in Northwest Mexico. *American Anthropologist* n.s. 37:287–305.

1943   The Chihuahua Culture Area. *New Mexico Anthropologist* 6–7(3):115–158.

Braniff, C. B.

1986   Ojo de Agua, Sonora, and Casas Grandes, Chihuahua: A Suggested Chronology. In *Ripples in the Chichimec Sea: New Considerations of Southwestern-Mesoamerican Interactions,* edited by F. J. Mathien and R. H. McGuire, pp. 70–80. Center for Archaeological Investigations and Southern Illinois University Press, Carbondale.

1988   A Propósito de el Ulama en el Norte de Mèxico. *Arqueologia* 3:47–94. Dirección de Monumentos Prehispánicos. Instituto Nacional de Antropología e Historia, México, D.F.

Burger, R. L.

1992   *Chavin and the Origins of Andean Civilizations*. Thames and Hudson, London.

Caldwell, J. R.

1964   Interaction Spheres in Prehistory. In *Hopewellian Studies,* edited by J. R. Caldwell and Bob Hall, pp. 133–143. Illinois State Museum Scientific Papers No. 12, Springfield Illinois.

Carey, H. A.

1931   An Analysis of the Chihuahua Culture. *American Anthropologist* 33:325–374.

Carlson, R.

1982   The Polychrome Complexes. In *Southwestern Ceramics, A Comparative Review,* edited by A. Schroeder, pp. 201–234. *Arizona Archaeologist* No. 15, Phoenix.

Carpenter, J. P.
1985   The Obsidian Hydration Dating of the Joyce Well Site. Ms. on file (Separate P-1247), Laboratory of Anthropology, Museum of New Mexico, Santa Fe.
Carpenter, J., G. Sanchez, and E. Villalpando
1999   Preliminary Investigations at La Playa, Sonora, Mexico. *Archaeology Southwest* 13(1):6.
Carpenter, J., and E. Villalpando
2001   From Paleoindians to Pedro Infante: Raising Time to the Level of Explanation at La Playa, Sonora. Paper presented at the 66th Annual Meeting of the Society for American Archaeology, April 18–22, 2001, New Orleans, Louisiana.
Coe, D. M.
1981   Religion and the Rise of Mesoamerican States. In *The Transitions to Statehood in the New World,* edited by G. D. Jones, and R. R. Kauty, pp.151–171. Cambridge University Press, Cambridge, U.K.
1989   The Olmec Heartland: Evolution of Ideology. In *Regional Perspectives on the Olmec,* edited by R. J. Sharer, and D. C. Grove, pp. 68–82. School of American Research, Santa Fe, New Mexico, and Cambridge University Press, Cambridge, U.K.
Cordell, L. S., and F. Plog
1979   Escaping the Confines of Normative Thought: A Re-evaluation of Puebloan Prehistory. *American Antiquity* 44(3):405–429.
Cox, J. R., and E. Blinman
1996   NSEP Archaeomagnetic Dating: Procedures, Results, and Interpretations. In *Pipeline Archaeology 1990–1993: The El Paso Natural Gas North System Expansion Project, New Mexico and Arizona,* vol. 12, *Supporting Studies: Nonceramic Artifacts, Subsistence and Environmental Studies, and Chronometric Studies.* Western Cultural Resource Management, Inc., Farmington, New Mexico.
Creel, D. G.
1997   Interpreting the End of the Mimbres Classic. In *Prehistory of the Border,* edited by J. Carpenter and G. Sanchez, pp. 25–31. Arizona State Museum Archaeological Series 186. University of Arizona, Tucson.
Crown, P. L.
1994   *Ceramics and Ideology: Salado Polychrome Pottery.* University of New Mexico Press, Albuquerque.
Dean, J. S., and J. C. Ravesloot
1993   The Chronology of Cultural Interaction in the Gran Chichimeca. In *Culture and Contact,* edited by A. I. Woosley and J. C. Ravesloot, pp. 83–103. Amerind Foundation Publication, Dragoon, and University of New Mexico Press, Albuquerque.
DeAtley, S. P.
1980   Regional Integration on the Northern Casas Grandes Frontier. Unpublished Ph.D. dissertation, Department of Anthropology, University of California, Los Angeles. University Microfilms, Ann Arbor.
DeAtley, S., and F. J. Findlow
1980   A Tentative Hydration Rate for the Antelope Wells Obsidian. *North American Archaeologist* 1(2).
1982   Regional Integration of the Northern Casas Grandes Frontier. In *Mogollon Archaeology: Proceedings of the 1980 Mogollon Conference,* edited by P. H. Beckett and K. Silverbird, pp. 263–277. Acoma Books, Ramona, California.
Di Peso, C. C.
1951   *The Babocomari Village Site on the Babocomari River, Southeastern Arizona.* Amerind Foundation Publication 5. Dragoon, Arizona.
1953   *The Sobaipuri Indians of the Upper San Pedro Valley, Southeastern Arizona.* Amerind Foundation Publication 6. Dragoon, Arizona.

1956   *The Upper Pima of San Cayetano del Tumacacori: An Archaeological Recon-struction of the Ootam of Pimeria Alta.* Amerind Foundation Publication 7. Dragoon, Arizona.

1958   *The Reeve Ruin of Southeastern Arizona.* The Amerind Foundation, Dragoon, Arizona.

1968   Casas Grandes and the Gran Chichimeca. *El Palacio* 75:45–61.

1974   *Casas Grandes: A Fallen Trade Center of the Gran Chichimeca*, vols. 1–3. Amerind Foundation, Dragoon, Arizona, and Northland Press, Flagstaff.

1979   Roots of the New Tradition: Prehistory of the Casas Grandes Valley. In *Juan Quezada and the New Tradition,* edited by D. Frankel, pp. 10–21. California State University Art Gallery, Fullerton.

1983   The Northern Sector of the Mesoamerican World System. In *Forgotten Places and Things: Archaeological Perspectives on American History,* edited by A. E. Ward, pp. 11–21. Contributions to Anthropological Studies 3. Center for Anthropological Research, Albuquerque.

Di Peso, C. C., J. B. Rinaldo, and G. J. Fenner

1974   *Casas Grandes: A Fallen Trade Center of the Gran Chichimeca*, vols. 4–8. Amerind Foundation, Dragoon, Arizona, and Northland Press, Flagstaff.

Dixon, K. A.

1963   The Interamerican Diffusion of a Cooking Technique: The Culinary Shoe-Pot. *American Anthropologist* 65:593–619.

1976   Shoe-Pots, Patajos, and the Principal of Whimsy. *American Antiquity* 41:386–391.

Douglas, J.

1987   Late Prehistoric Archaeological Remains in the San Bernardino Valley, Southeastern Arizona. *The Kiva* 53:35–51.

1992   Distant Sources, Local Contexts: Interpreting Nonlocal Ceramics at Paquimé (Casas Grandes) Chihuahua. *Journal of Anthropological Research* 48(1):1–24.

1995   Autonomy and Regional Systems in the Late Prehistoric Southern Southwest. *American Antiquity* 60:240–257.

Doyel, D. E.

1976   Salado Cultural Development in the Tonto Basin and Globe-Miami Areas, Central Arizona. *The Kiva* 42(1):5–16.

1994   Charles Corrandino Di Peso: Expanding the Frontiers of New World Prehistory. *American Antiquity* 59:9–20.

DuBois, R. L.

1989   Archaeomagnetic Results from the Southwest United States and Mesoamerica, and Comparison with Some Other Areas. *Physics of the Earth and Planetary Interiors* 56:18–23.

Duran, M. S.

1992   Animas Phase Sites in Hidalgo County, New Mexico. National Register of Historic Places, Multiple Property Documentation Form. Prepared for the New Mexico Historic Preservation Division, Santa Fe.

Fewkes, J. W.

1892   The Ceremonial Circuit among the Village Indians of Northeastern Arizona. *Journal of American Folk-Lore* 5(16):33–42.

Findlow, F. J., and S. P. DeAtley

1978   An Ecological Analysis of Animas Phase Assemblages in Southwestern New Mexico. *Journal of New World Archaeology* 2(5):5–18.

Fish, P. R., and S. K. Fish

1999   Reflections on the Casas Grandes Regional System from the Northwestern Periphery. In *The Casas Grandes World,* edited by C. F. Schaafsma and C. L. Riley, pp. 27–42. University of Utah Press, Salt Lake City.

Freidel, D., and L. Schele
1988   Dead Kings and Living Temples: Dedication and Termination Rituals among the Ancient Maya. In *Word and Image in Maya Culture: Explorations in Language, Writing, and Representation,* edited by W. F. Hanks and D. S. Rice, pp. 233–243. University of Utah Press, Salt Lake City.

Gillespie, S. D.
1991   Ballgames and Boundaries. In *The Mesoamerican Ballgame,* edited by V. L. Scarborough and D. R. Wilcox, pp. 317–346. University of Arizona Press, Tucson.

Gladwin, W., and H. S. Gladwin
1930   *Some Southwestern Pottery Types: Series I.* Medallion Papers No. 25. Gila Pueblo, Globe, Arizona.

1934   *A Method for the Designation of Cultures and Their Variations.* Medallion Papers No. 15. Gila Pueblo, Globe, Arizona.

Grieder, T., A. B. Mendoza, C. E. Smith Jr., and R. M. Marina
1988   *La Galgada.* University of Texas Press, Austin.

Hill, J. D.
1995   *Ritual and Rubbish in the Iron Age of Wessex: A Study on the Formation of A Specific Archaeological Record.* BAR Series 242. British Archaeological Reports, Oxford, U.K.

Hooten, E. A.
1930   *The Indians of Pecos Pueblo.* New Haven, Connecticut.

Johnson, A. E., and R. H. Thompson
1963   The Ringo Site, Southeastern Arizona. *American Antiquity* 28:465–481.

Keatinge, R. W.
1981   The Nature and Role of Religious Diffusion in the Early Stages of State Formation: An Example from Peruvian Prehistory. In *The Transition to Statehood in the New World,* edited by G. D. Jones and R. R. Kautz, pp. 172–187. Cambridge University Press, Cambridge, U.K.

Kelley, J. C.
1948   Jumano and Patarabueye: Relations at La Junta de los Rios. Unpublished Ph.D. dissertation, Department of Anthropology, Harvard University, Cambridge.

1951   A Bravo Valley Aspect Component of the Lower Conchos River Valley, Chihuahua, Mexico. *American Antiquity* 17(2):114–119.

1966   Mesoamerica and the Southwestern United States. In *Archaeological Frontiers and External Connections,* edited by G. F. Ekholm and G. R. Willey, pp. 95–110. Handbook of Middle American Indians, vol. 11. University of Texas Press, Austin.

1991   The Known Ballcourts of Durango and Zacatecas. In *The Mesoamerican Ballgame,* edited by V. L. Scarborough and D. R. Wilcox, pp. 87–101. University of Arizona Press, Tucson.

Kelley, J. H., J. D. Stewart, A. C. MacWilliams, and L. C. Neff
1999   A West Central Chihuahuan Perspective on Chihuahuan Culture. In *The Casas Grandes World,* edited by C. F. Schaafsma and C. L. Riley, pp. 63–77. University of Utah Press, Salt Lake City.

Kent, K. P.
1957   The Cultivation and Weaving of Cotton in the Prehistoric Southwestern United States. *Transactions of the American Philosophical Society* 47(3):457–732. Philadelphia, Pennsylvania.

Kidder, A. V., H. S. Cosgrove, and C. B. Cosgrove
1949   The Pendleton Ruin, Hidalgo County, New Mexico. Contributions to American Anthropology and History 10. *Carnegie Institution of Washington Publication* 585:107–152. Washington, D.C.

Kowalewski, S. A., G. M. Feinman, L. Finster, and R. Blanton
1991   Pre-Hispanic Ballcourts from the Valley of Oaxaca, Mexico. In *The Mesoamer-*

ican Ballgame, edited by V. L. Scarborough and D. R. Wilcox, pp. 25–44. University of Arizona Press, Tucson.

LaBelle, J. M., and J. L. Eighmy

1995   *1995 Additions to the List of Independently Dated Virtual Geomagnetic Poles and the Southwest Master Curve.* CSU Archaeomagnetic Lab Technical Series 7. California State University, Santa Barbara.

Lambert, M. F., and J. R. Ambler

1965   *A Survey and Excavation of Caves in Hildago County, New Mexico.* School of American Research Monograph No. 25. School of American Research, Santa Fe, New Mexico.

LaMotta, V. M., and M. B. Schiffer

1999   Formation Processes of House Floor Assemblages. In *The Archaeology of Household Activities,* edited by P. Allison. Routledge, London.

LeBlanc, S. A.

1975   Mimbres Archaeological Center: Preliminary Report of the First Season of Excavation, 1974. Institute of Archaeology, University of California, Los Angeles.

1977   The 1976 Field Season of the Mimbres Foundation in Southwestern New Mexico. *Journal of New World Archaeology* 2(2).

1980a   The Post Mogollon Periods in Southwestern New Mexico: The Animas/Black Mountain Phase and the Salado Period. In *An Archaeological Synthesis of South-Central and Southwestern New Mexico,* edited by S. A. LeBlanc and M. E. Whalen, pp. 271–316. Office of Contract Archeology, University of New Mexico, Albuquerque.

1980b   The Dating of Casas Grandes. *American Antiquity* 45(4):799–806.

1999   *Prehistoric Warfare in the American Southwest.* University of Utah Press, Salt Lake City.

LeBlanc, S. A., and B. Nelson

1976   The Salado in Southwestern New Mexico. *The Kiva* 42(1):71–79.

Lehmer, D.

1948   *The Jornada Branch of the Mogollon.* Social Sciences Bulletin 17, University of Arizona Bulletin 19(2). Tucson.

Lekson, S.

1984   Dating Casas Grandes. *The Kiva* 50(1):55–60.

1999   *Chaco Meridian: Centers of Political Power in the Ancient Southwest.* Altamira Press, Walnut Creek, California.

2000   Salado in Chihuahua. In *Salado,* edited by J. S. Dean. Amerind Foundation, Dragoon, Arizona, and University of New Mexico Press, Albuquerque.

Lekson, S., and T. Klinger

1973   Villareal II: Preliminary Notes on an Animas Phase Site in Southwestern New Mexico. *Awanyu* 1(2).

Leyenaar, T.J.J.

1992   *Ulama.* The Survival of the Mesoamerican Ballgame *Ullamaliztli. Kiva* 58:115–154.

Lightfoot, R. R.

1993   Synthesis. In *The Duckfoot Site,* vol. 1, *Descriptive Archaeology,* edited by R. R. Lightfoot and M. C. Etzkorn, pp. 297–302. Occasional Paper No. 3, Crow Canyon Archaeological Center, Cortez, Colorado.

Lister, R. H.

1938   Some Aspects of Chihuahua Archaeology. Unpublished Master's thesis, Department of Anthropology, University of New Mexico, Albuquerque. Ms. on file, Department of Anthropology, University of New Mexico.

1946   Survey of Archaeological Remains in Northwestern Chihuahua. *Southwest Journal of Anthropology* 2:433–453.

1955   *The Present Status of the Archaeology of Western Mexico: A Distributional Study.* University of Colorado Studies in Anthropology, No. 5. Boulder.

Long, A., and B. Rippeteau
1974   Testing Contemporaneity and Averaging of Radiocarbon Dates. *American Antiquity* 39:205–215.

Lowdon, J. A.
1969   Isotopic Fractionation in Corn. *Radiocarbon* 11(2):391–393.

McCluney, E. B.
1965a   *Clanton Draw and Box Canyon: An Interim Report on Two Prehistoric Sites in Hidalgo County, New Mexico, and Related Surveys.* School of American Research Monograph No. 26. Santa Fe, New Mexico.
1965b   The Excavation of the Joyce Well Site, Hidalgo County, New Mexico. Ms. on file at the School of American Research and at the Laboratory of Anthropology, Museum of New Mexico, Santa Fe.

McGuire, R. H.
1980   The Mesoamerican Connection in the Southwest. *The Kiva* 46:3–38.
1986   Economies and Modes of Production in the Prehistoric Southwestern Periphery. In *Ripples in the Chichimec Sea: New Considerations of Southwestern-Mesoamerican Interactions,* edited by F. Mathien and R. McGuire, pp. 243–269. Southern Illinois University Press, Carbondale.
1989   The Greater Southwest as a Periphery of Mesoamerica. In *Centre and Periphery,* edited by T. C. Champion, pp. 40–66. Allen and Unwin, London.

McGuire, R. H., M. E. Villalpando, V. D. Vargas, and E. Gallaga
1999   Cerro de Trincheras and the Casas Grandes World. In *The Casas Grandes World,* edited by Curtis F. Schaafsma and Carroll L. Riley, pp. 134–146. University of Utah Press, Salt Lake City.

Michels, J. W., I.S.T. Tsong, and G. A. Smith
1983   Experimentally Derived Hydration Rates in Obsidian Dating. *Archaeometry* 25:107–117.

Minnis, P. E.
1984   Peeking Under the Tortilla Curtain: Regional Interaction and Integration on the Northern Periphery of Casas Grandes. *American Archaeology* 4:181–193.
1988   Four Examples of Specialized Production at Casas Grandes, Northwest Chihuahua. *The Kiva* 53:181–193.
1989   The Casas Grandes Polity in the International Four Corners. In *The Sociopolitical Structure of Prehistoric Southwestern Societies,* edited by S. Upham, K. Lightfoot, and R. Jewett, pp. 269–305. Westview Press, Boulder, Colorado.

Minnis, P., and M. Whalen
1990   El Sistema Regional de Casas Grandes, Chihuahua. In *Actas del Segundo Congreso de Historia Regional Comparada,* pp. 45–55. Universidad Autonoma de Ciudad Juarez, Chihuahua, Mexico.

Mock, S. B. (editor)
1998   *The Sowing and the Dawning: Termination, Dedication, and Transformation in the Archaeological and Ethnographic Record of Mesoamerica.* University of New Mexico Press, Albuquerque.

Montgomery, B.
1993   Ceramic Analysis as a Tool for Discovering Processes of Pueblo Abandonment. In *Abandonment of Sites and Regions,* edited by C. M. Cameron and S. A. Tomka, pp. 157–164. Cambridge University Press, Cambridge, U.K.

Moore-Craig, Narca
1996   The Historical, Cultural, and Ecological Overview of the Gray Ranch. Ms. on file, Animas Foundation, Animas, New Mexico.

Naylor, T. H.
1985   Casas Grandes Outlier Ballcourts in Northwest Chihuahua. Paper presented at the International Mesoamerican Ballcourt Symposium, Tucson.

1995   Casas Grandes Outlier Ball Courts in Northwest Chihuahua. In *The Gran Chichimeca: Essays on the Archaeology and History of Northern Mesoamerica,* edited by J. E. Reyman, pp. 224–239. Ashgate, Brookfield, Vermont.

Nickerson, N. H.

1953   Variation in Cob Morphology among Certain Archaeological and Ethnological Races of Maize. *Annals of the Missouri Botanical Garden* 40(2):79–111. St. Louis.

Nielsen, A. E.

1991   Trampling the Archaeological Record: An Experimental Study. *American Antiquity* 56:483–503.

Northrop, S. A.

1959   *Minerals of New Mexico*. University of New Mexico Press, Albuquerque.

Ortiz, A.

1972   *The Tewa World*. University of Chicago Press, Chicago, Illinois.

Pailes, R. A., and D. T. Reff

1985   Colonial Exchange Systems and the Decline of Paquimé. In *The Archaeology of West and Northwest Mesoamerica,* edited by M. S. Foster and P. C. Weigand, pp. 353–364. Westview Press, Boulder, Colorado.

Parsons, E. C.

1996   *Pueblo Indian Religion*. University of Nebraska Press, Lincoln (originally published in 1939).

Pasztory, E.

1972   The Historical and Religious Significance of the Middle Classic Ball Game. In *Religion en Mesoamerica,* edited by J. L. King and N. Castillo Tejero, pp. 441–455. Sociedad Mexican de Antropologia, Mexico.

1976   *The Murals of Tepantitla, Teotihuacan*. Garland Press, New York.

1978   Artistic Traditions of the Middle Classic Period. In *Middle Classic Mesoamerican: A.D. 400–700,* edited by E. Pasztory, pp. 108–142. Columbia University Press, New York.

Phillips, D. A., Jr., and J. P. Carpenter

1999   A Re-evaluation of the Robles Phase of the Casas Grandes Culture, Northwest Chihuahua. In *The Casas Grandes World,* edited by C. F. Schaafsma and C. L. Riley, pp. 78–83. University of Utah Press, Salt Lake City.

Phillips, D. A., Jr., and C. F. Schaafsma

1987   Archaeology at the Museum of New Mexico—Past, Present, Future. *El Palacio* 93:38–41.

Plog, F.

1984   Exchange, Tribes, and Alliances: The Northern Southwest. *American Archaeologist* 4(3):412–423.

1979   Prehistory: Western Anasazi. In *Handbook of North American Indians,* vol. 9, *Southwest,* edited by A. Ortiz, pp. 108–130. Smithsonian Institution, Washington, D.C.

Plog, F., S. Upham, and P. Weigand

1982   A Perspective on Mogollon-Mesoamerican Interaction. In *Mogollon Archaeology: Proceedings of the 1980 Mogollon Conference,* edited by P. Beckett, pp. 227–247. Acoma Books, Ramona, California.

Plog, S.

1980   *Stylistic Variation of Prehistoric Ceramics*. Cambridge University Press, Cambridge, U.K.

Pryor, F.

1977   *The Origins of the Economy: A Comparative Study of Distribution in Primitive and Peasant Economies*. Academic Press, New York.

Ravesloot, J. C.

1979   The Animas Phase: Post Classic Mimbres Occupation of the Mimbres Valley,

New Mexico. M.A. thesis, Department of Anthropology, Southern Illinois University, Carbondale.

1988   *Mortuary Practices and Social Differentiation at Casas Grandes, Chihuahua, Mexico.* Anthropological Papers of the University of Arizona No. 49. University of Arizona Press, Tucson.

Ravesloot, J. C., J. S. Dean, and M. S. Foster

1986   A New Perspective on Casas Grandes Tree-Ring Dates. Paper presented at the 4th Mogollon Conference, University of Arizona, Tucson.

1995   New Perspective on the Casas Grandes Tree-Ring Dates. In *The Gran Chichimeca: Essays on the Archaeology and History of Northern Mesoamerica,* edited by J. E. Reyman, pp. 240–251. Ashgate, Brookfield, Vermont.

Ravesloot, J., and M. Foster

1984   National Register of Historic Places Inventory Nomination Form, Site LA-54049. On file at the Museum of New Mexico, Santa Fe.

Ravesloot, J., and P. Spoerl

1987   The Role of Warfare in the Development of Status Hierarchies at Casas Grandes, Chihuahua, Mexico. Paper presented at the 20th annual Chacmool Conference, University of Calgary, Alberta, Canada.

1989   The Role of Warfare in the Development of Status Hierarchies at Casas Grandes, Chihuahua, Mexico. In *Cultures in Conflict: Current Archaeological Perspectives,* edited by D. C. Tkaczuk and B. C. Vivian, pp. 130–137. University of Calgary Archaeological Association, Calgary.

Reed, E. K.

1949   The Significance of Skull Deformation in the Southwest. *El Palacio* 56(4):106–119.

1963   *Occipital Deformation in the Northern Southwest.* Regional Research Abstract No. 310. National Park Service, Santa Fe, New Mexico.

n.d.   Human Skeletal Material from Galisteo Basin Ruins. Report for the Museum of New Mexico (in preparation in 1965).

Renfrew, C., and J. Cherry

1986   *Peer Polity Interaction and Socio-Political Change.* Cambridge University Press, Cambridge, U.K.

Riley, C.

1987   *The Frontier People: The Greater Southwest in the Proto-Historic Period.* University of New Mexico Press, Albuquerque.

Santley, R. S., M. J. Berman, and R. T. Alexander

1991   The Politicization of the Mesoamerican Ballgame and its Implications for the Interpretation of the Distribution of Ball Courts in Central Mexico. In *The Mesoamerican Ballgame,* edited by V. L. Scarborough and D. R. Wilcox, pp. 3–24. University of Arizona Press, Tucson.

Sauer, C., and D. Brand

1930   Pueblo Sites in Southeastern Arizona. *University of California Publications in Geography* 3(7):415–458.

Sayles, E. B.

1936a   *An Archaeological Survey of Chihuahua, Mexico.* Medallion Papers No. 22. Gila Pueblo, Globe, Arizona.

1936b   *Some Southwestern Pottery Types.* Medallion Papers No. 21. Gila Pueblo, Globe, Arizona.

Schaafsma, C. F.

1979   The "El Paso Phase" and its Relationship to the "Casas Grandes Phenomenon." In *Jornada Mogollon Archaeology: Proceedings of the First Jornada Conference,* edited by P. H. Beckett and R. N. Wiseman, pp. 383–388. New Mexico State University, Las Cruces, and Historic Preservation Bureau, Santa Fe, New Mexico.

1987   Statement of Curtis F. Schaafsma, New Mexico State Archaeologist and President, American Society for Conservation Archaeology, before the Committee on Interior and Insular Affairs: Oversight Hearing on Phoenix Indian School Property Disposition. *American Society for Conservation Archaeology Report* 14(2):7–11. On file, Laboratory of Anthropology Library, Santa Fe, New Mexico.

1995   The Casas Grandes Interaction Sphere: Origins, Nature, Contacts and Legacy. Paper presented at the Durango Conference on Southwest Archaeology. Durango, Colorado, September 16.

1997   Ethnohistoric Groups in the Casas Grandes Interaction Sphere: Circa A.D. 1500–1700. In *Layers of Time: Papers in Honor of Robert H. Weber,* edited by M. S. Duran and D. T. Kirkpatrick, pp. 85–98. Archaeological Society of New Mexico, Albuquerque.

Schaafsma, C. F., and C. L. Riley (editors)

1999   *The Casas Grandes World.* University of Utah Press, Salt Lake City.

Schaafsma, C. F., and D. Wolfman

1989   Recent Archaeomagnetic Dates for the Joyce Well Site in Hidalgo County, New Mexico and the Implications for Dating the Animas Phase and the Related Casas Grandes Culture. Paper presented at the 1989 annual meeting of the Archaeological Society of New Mexico, Taos. On file, Laboratory of Anthropology Library, Santa Fe, New Mexico.

Schaafsma, P.

1997   *Rock Art Sites in Chihuahua, Mexico.* Archaeology Notes 171. Office of Archaeological Studies, Museum of New Mexico, Santa Fe.

1998   The Paquimé Rock Art Style, Chihuahua, Mexico. In *Rock Art of the Chihuahuan Desert Borderlands,* edited by S. Smith-Savage and R. J. Mallouf, pp. 33–44. Center for Big Bend Studies, Sul Ross State University, Alpine, Texas.

Schiffer, M. B.

1972   Archaeological Context and Systemic Context. *American Antiquity* 37:156–165.

1987   *Formation Processes of the Archaeological Record.* University of New Mexico Press, Albuquerque.

Schiffer, M. B., and J. M. Skibo

1987   Theory and Experiment in the Study of Technological Change. *Current Anthropology* 28:595–622.

1997   The Explanation of Artifact Variability. *American Antiquity* 62:27–50.

Schortman, E. M., and P. A. Urban

1987   Modeling Interregional Interaction in Prehistory. In *Advances in Archaeological Method and Theory,* vol. 11, edited by M. B. Schiffer, pp. 37–95. Academic Press, New York.

Schwennesen, A. T.

1918   *Ground Water in Animas, Playas, Hachita and San Luis Basins, New Mexico.* U.S. Geological Survey, Water Supply Papers, No. 422. Government Printing Office, Washington, D. C.

Sebastian, L.

1992   *The Chaco Anasazi: Sociopolitical Evolution in the Prehistoric Southwest.* Cambridge University Press, New York.

Seltzer, C. C.

1944   *Racial Prehistory in the Southwest and the Hawikuh Zunis.* Peabody Museum Papers vol. 23, no. 1. Cambridge, Massachusetts.

Seymour, D. J., and M. B. Schiffer

1987   A Preliminary Analysis of Pithouse Assemblages from Snaketown, Arizona. In *Method and Theory for Activity Area Research: An Ethnoarchaeological Approach,* edited by S. Kent, pp. 549–603. Columbia University Press, New York.

Shafer, H. J.
1999   The Mimbres Classic and Postclassic. In *The Casas Grandes World,* edited by Curtis F. Schaafsma and Carrol L. Riley, pp. 121–133. University of Utah Press, Salt Lake City.

Skibo, J. M., and M. B. Schiffer
2001   Understanding Artifact Variability and Change: A Behavioral Framework. In *The Anthropology of Technology,* edited by M. B. Schiffer, pp. 139–150. Amerind Foundation, Dragoon, Arizona, and the University of New Mexico Press, Albuquerque.

Skibo, J. M., M. B. Schiffer, and K. C. Reid
1989   Organic-Tempered Pottery: An Experimental Study. *American Antiquity* 54:122–146.

Stallings, W. S., Jr.
1931   *El Paso Polychrome.* Technical Series Bulletin, No. 3. Laboratory of Anthropology, Santa Fe, New Mexico.

Stevanovic, M.
1997   The Age of Clay: The Social Dynamics of House Destruction. *Journal of Anthropological Archaeology* 16:334–395.

Stevenson, C., J. Carpenter, and B. Scheetz
1989   Obsidian Dating: Recent Advances in the Experimental Determination and Application of Obsidian Hydration Rates. *Archaeometry* 31(2):193–206.

Stubbs, S. A., and W. S. Stalling Jr.
1953   *The Excavation of Pindi Pueblo, New Mexico.* Monographs of the School of American Research 18. Santa Fe, New Mexico.

Sullivan, A. P.
1978   Inference and Evidence: A Discussion of the Conceptual Problems. *Advances in Archaeological Method and Theory,* vol. 1, edited by M. B. Schiffer, pp. 183–222. Academic Press, New York.

Sullivan, A. P., and K. C. Rozen
1985   Debitage Analysis and Archaeological Interpretation. *American Antiquity* 50:755–779.

Turner, V.
1974   *Dramas, Fields, and Metaphors: Symbolic Action in Human Society.* Cornell University Press, Ithaca, New York.

Turner, C. G. II, and J. A. Turner
1998   *Cannibalism and Violence in the Prehistoric American Southwest.* University of Utah Press, Salt Lake City.

Upham, S.
1982   *Polities and Power: An Economic and Political History of the Western Pueblo.* Academic Press, New York.

Upham, S., K. Lightfoot, and G. Feinman
1981   Explaining Socially Determined Ceramic Distributions in the Prehistoric Plateau Southwest. *American Antiquity* 46:822–833.

VanPool, T. L., C. S. VanPool, R. C. Cruz Antillón, R. D. Leonard, and M. J. Harmon
2000   Flaked Stone and Social Interaction in the Casas Grandes Region, Chihuahua, Mexico. *Latin American Antiquity* 11:163–174.

Vargas, V. D.
1995   *Copper Bell Trade Patterns in the Prehistoric U. S. Southwest and Northwest Mexico.* Arizona State Museum, Archaeological Series 187. Arizona State Museum, University of Arizona, Tucson.

Walker, W. H.
1995a   Ceremonial Trash? In *Expanding Archaeology,* edited by J. M. Skibo, W. H. Walker, and A. E. Nielsen, pp. 67–79. University of Utah Press, Salt Lake City.

1995b   Ritual Prehistory: A Pueblo Case Study. Unpublished Doctoral Dissertation, Department of Anthropology, University of Arizona.

1996   Ritual Deposits: Another Perspective. In *River of Change: Prehistory of the Middle Little Colorado River Valley, Arizona*, edited by E. C. Adams, pp. 75–91. Arizona State Museum Archaeological Series No. 185, University, Tucson.

1998   Where are the Witches of Prehistory? *Journal of Archaeological Method and Theory* 5(2):245–308.

2001   Ritual Technology in an Extranatural World. In *Anthropological Perspectives on Technology*, edited by M. B. Schiffer, pp. 87–106. Amerind Foundation, Dragoon, Arizona and the University of New Mexico Press, Albuquerque.

2002   Stratigraphy and Practical Reason. *American Anthropologist* 104:159–177.

Wallerstein, I.

1974   *The Modern World System.* Academic Press, New York.

Werbner, R. P.

1977   Introduction. In *Regional Cults,* edited by R. P. Werbner, pp. ix–xxxvii. Academic Press, London.

Whalen, M.

1978   *Settlement Patterns in the Western Hueco Bolson.* Anthropological Paper No. 6, El Paso Centennial Museum. El Paso, Texas.

Whalen, M. E., and P. E. Minnis

1996   Ball Courts and Political Centralization in the Casas Grandes Region. *American Antiquity* 61:732–746.

1999   Investigating the Paquimé Regional System. In *The Casas Grandes World,* edited by C. F. Schaafsma and C. L. Riley, pp. 54–62. University of Utah Press, Salt Lake City.

2001   *Casas Grandes and its Hinterland: Prehistoric Regional Organization in Northwest Mexico.* University of Arizona Press, Tucson.

Wheat, J. B.

1955   *Mogollon Culture prior to* A.D. *1000.* Memoir No. 82. American Anthropological Association, Menasha.

White, W.

1986   Dissolution Mechanisms of Nuclear Waste Glass: A Review. *Advances in Ceramics,* vol. 20. *Nuclear Waste Management* II:431–442.

Whitecotton, J., and R. Pailes

1986   New World Precolumbian World Systems. In *Ripples in the Chichimec Sea: New Considerations of Southwestern-Mesoamerican Interaction,* edited by F. Mathien and R. McGuire, pp. 183–204. Southern Illinois University Press, Carbondale.

Wilcox, D. R.

1991   The Mesoamerican Ballgame in the American Southwest. In *The Mesoamerican Ballgame,* edited by V. Scarborough and D. Wilcox, pp. 101–125. University of Arizona Press, Tucson.

1995   A Processual Model of Charles C. Di Peso's Babocomari Site and Related Systems. In *The Gran Chichimeca: Essays in on the Archaeology and Ethnohistory of Northern Mesoamerica,* edited by J. Reyman, pp. 281–319. Aldershot, Avebury.

Wilcox, D. R., and L. Shenk

1977   *The Architecture of the Casa Grande and Its Interpretation.* Arizona State Museum Archaeological Series No.115. University of Arizona, Tucson.

Wilcox, D. R., and C. Sternberg

1983   *Hohokam Ballcourts and Their Interpretation.* Arizona State Museum Archaeological Series No. 160. University of Arizona, Tucson.

Wilkerson, S. J.

1991   And Then They Were Sacrificed: The Ritual Ballgame of Northeastern

Mesoamerica Through Time and Space. In *The Mesoamerican Ballgame*, edited by V. L. Scarborough and D. R. Wilcox, pp. 45–72. University of Arizona Press, Tucson.

Willey, G. R.
1948   A Functional Analysis of Horizon Styles in Peruvian Archaeology. In *A Reappraisal of Peruvian Archaeology*, edited by W. C. Bennet, pp. 8–15. Memoir 4, Society for American Archaeology, Menasha.

Wilshusen, R. H.
1986   The Relationship between Abandonment Mode and Ritual Use in Pueblo I Anasazi Protokivas. *Journal of Field Archaeology* 13:245–254.
1988   The Abandonment of Structures. In *Dolores Archaeological Program: Supporting Studies: Additive and Reductive Technologies*, edited by E. R. Blinman, C. J. Phagan, and R. H. Wilshusen, pp. 673–702. Bureau of Reclamation, Engineering, and Research Center, Denver, Colorado.

Wilshusen, R. H., and S. G. Ortman
1999   Rethinking the Pueblo I Period in the San Juan Drainage: Aggregation, Migration, and Cultural Diversity. *Kiva* 64:369–399.

Wolf, E.
1982   *Europe and the People Without History*. University of California Press, Berkeley.

Wolfman, D. W.
1984   Geomagnetic Dating Methods in Archaeology. In *Advances in Archaeological Method and Theory*, vol. 7, edited by M. B. Schiffer, pp. 363–458. Academic Press, Orlando, Florida.
1990   Retrospect and Prospect. In *Archaeomagnetic Dating*, edited by J. L. Eighmy and R. S. Sternberg, pp. 313–364. University of Arizona Press, Tucson.

Wolfman, D. W., and C. F. Schaafsma
1989   Recent Archaeomagnetic Dates for the Joyce Well Site in Hidalgo County, New Mexico and the Implications for Dating the Animas Phase and the Related Casas Grandes Culture. Paper presented at the 1989 annual meeting of the Archaeological Society of New Mexico, Taos.

Woosley, A., and B. Olinger
1993   The Casas Grandes Ceramic Tradition: Production and Interregional Exchange of Ramos Polychrome. In *Culture and Contact: Charles Di Peso's Gran Chichimeca*, edited by A. I. Woosley and J. C. Ravesloot, pp. 105–131. Amerind Foundation, Dragoon, and University of New Mexico Press, Albuquerque.

Yoffee, N.
1979   The Decline and Rise of Mesopotamian Civilization. *American Antiquity* 44:5–35.

Zeller, R. A.
1962   *Reconnaissance Geologic Map of Southern Animas Mountains*. State Bureau of Mines and Mineral Resources, New Mexico Institute of Mining and Technology, Socorro.
1965   *Stratigraphy of the Big Hatchet Mountain area, New Mexico*. Memoir No. 16, State Bureau of Mine and Mineral Resources. New Mexico Institute of Mining and Technology, Socorro.

Zeller, R. A., and A. M. Alper
1965   *Geology of the Walnut Wells Quadrangle, Hidalgo County, New Mexico*. State Bureau of Mines and Mineral Resources, New Mexico Institute of Mining and Technology, Socorro.

INDEX